Interface between Quantum Information and Statistical Physics

Kinki University Series on Quantum Computing

Editor-in-Chief: Mikio Nakahara *(Kinki University, Japan)*

ISSN: 1793-7299

Published

Vol. 1 Mathematical Aspects of Quantum Computing 2007
edited by Mikio Nakahara, Robabeh Rahimi (Kinki Univ., Japan) &
Akira SaiToh (Osaka Univ., Japan)

Vol. 2 Molecular Realizations of Quantum Computing 2007
edited by Mikio Nakahara, Yukihiro Ota, Robabeh Rahimi, Yasushi Kondo &
Masahito Tada-Umezaki (Kinki Univ., Japan)

Vol. 3 Decoherence Suppression in Quantum Systems 2008
edited by Mikio Nakahara, Robabeh Rahimi & Akira SaiToh
(Kinki Univ., Japan)

Vol. 4 Frontiers in Quantum Information Research: Decoherence, Entanglement, Entropy, MPS and DMRG 2009
edited by Mikio Nakahara (Kinki Univ., Japan) &
Shu Tanaka (Univ. of Tokyo, Japan)

Vol. 5 Diversities in Quantum Computation and Quantum Information
edited by edited by Mikio Nakahara, Yidun Wan & Yoshitaka Sasaki
(Kinki Univ., Japan)

Vol. 6 Quantum Information and Quantum Computing
edited by Mikio Nakahara & Yoshitaka Sasaki (Kinki Univ., Japan)

Vol. 7 Interface Between Quantum Information and Statistical Physics
edited by Mikio Nakahara (Kinki Univ., Japan) &
Shu Tanaka (Univ. of Tokyo, Japan)

Vol. 8 Lectures on Quantum Computing, Thermodynamics and Statistical Physics
edited by Mikio Nakahara (Kinki Univ., Japan) &
Shu Tanaka (Univ. of Tokyo, Japan)

Kinki University Series on Quantum Computing – Vol. 7

editors

Mikio Nakahara
Kinki University, Japan

Shu Tanaka
University of Tokyo, Japan

Interface between Quantum Information and Statistical Physics

World Scientific

NEW JERSEY · LONDON · SINGAPORE · BEIJING · SHANGHAI · HONG KONG · TAIPEI · CHENNAI

Published by

World Scientific Publishing Co. Pte. Ltd.
5 Toh Tuck Link, Singapore 596224
USA office: 27 Warren Street, Suite 401-402, Hackensack, NJ 07601
UK office: 57 Shelton Street, Covent Garden, London WC2H 9HE

British Library Cataloguing-in-Publication Data
A catalogue record for this book is available from the British Library.

Kinki University Series on Quantum Computing — Vol. 7
INTERFACE BETWEEN QUANTUM INFORMATION AND STATISTICAL PHYSICS

Copyright © 2013 by World Scientific Publishing Co. Pte. Ltd.

All rights reserved. This book, or parts thereof, may not be reproduced in any form or by any means, electronic or mechanical, including photocopying, recording or any information storage and retrieval system now known or to be invented, without written permission from the Publisher.

For photocopying of material in this volume, please pay a copying fee through the Copyright Clearance Center, Inc., 222 Rosewood Drive, Danvers, MA 01923, USA. In this case permission to photocopy is not required from the publisher.

ISBN 978-981-4425-27-8

Printed in Singapore.

PREFACE

This volume contains lecture notes presented at the Symposium on Interface between Quantum Information and Statistical Physics, held from 10 to 12 November, 2011 at Kinki University, Osaka, Japan. The aim of the symposium was to exchange and share ideas among researchers working in various fields related to quantum information theory and statistical physics. Lecturers were asked to make their presentations accessible to other researchers with various backgrounds. Articles in this volume reflect this request and they should be accessible not only to forefront researchers but also to beginning graduate students or advanced undergraduate students.

This symposium was supported financially by "Open Research Center" Project for Private Universities: matching fund subsidy from MEXT (Ministry of Education, Culture, Sports, Science and Technology).

We would like to thank all the lecturers and participants, who made this symposium invaluable. We would like to thank Shoko Kojima for her dedicated secretarial work and Takumi Nitanda and Ryota Mizuno for TeXnical assistance. Finally, we would like to thank Zhang Ji and Rhaimie B Wahap of World Scientific for their excellent editorial work.

Mikio Nakahara
Shu Tanaka

Osaka and Tokyo, February 2012

SYMPOSIUM ON INTERFACE BETWEEN QUANTUM INFORMATION AND STATISTICAL PHYSICS

Kinki University (Osaka, Japan)

10 – 12 November 2011

10 November 2011

[Chair: Shu Tanaka]
Kenichi Kasamatsu (Kinki University)
 Bosons in an Optical Lattice with a Synthetic Magnetic Field

Tim Byrnes (NII)
 Quantum Simulation Using Exciton-Polaritons and Their Applications Toward Accelerated Optimization Problem Search

[Chair: Tasheng Tai]
Seiji Sugawa (Kyoto University)
 Quantum Simulation Using Ultracold Atoms in Optical Lattices

Akira Saitoh (Kinki University)
 Progress in a Quantum Genetic Algorithm

11 November 2011

[Chair: Mikio Nakahara]
Tasheng Tai (Kinki University)
 Universality of Integrable Model

Kohei Motegi (Okayama Institute for Quantum Physics)
 Exact Analysis of Correlation Functions of the XXZ Chain

[Chair: Ryo Tamura]
Hiroyuki Tomita (Kinki University)
 Classical Analogue of Weak Value in Stochastic Process

Hiroaki Matsueda (Sendai National College of Technology)
 Scaling of Entanglement Entropy and Hyperbolic Geometry

[Chair: Jun-ichi Inoue]
Brunello Tirozzi (University of Rome)
 Quantum Neural Networks, An Immersion of Classical Neural Networks in the Domain of Quantum Computing

12 November 2011

[Chair: Hiroaki Matsueda]
Jun-ichi Inoue (Hokkaido University)
 Analysis of Quantum Monte Carlo Dynamics in Infinite-Range Ising Spin Systems: Theory and Its Possible Applications

Ryo Tamura (University of Tokyo)
 A Method to Control Order of Phase Transition: Invisible States in Discrete Spin Models

Shu Tanaka (University of Tokyo)
 Some Topics in Quantum Annealing

LIST OF PARTICIPANTS

Brunello, Tirozzi	University of Rome, Italy
Chiara, Bagnasco	Kinki University, Japan
Fujimoto, Masahumi	Nara Medical University, Japan
Ichikawa, Tsubasa	Kinki University, Japan
Inoue, Jun	Hokkaido University, Japan
Kasamatsu, Kenichi	Kinki University, Japan
Kondo, Yasushi	Kinki University, Japan
Machide, Tomoya	Kinki University, Japan
Matsueda, Hiroaki	Sendai National College of Technology, Japan
Motegi, Kohei	Okayama Institute for Quantum Physics, Japan
Nakahara, Mikio	Kinki University, Japan
Obuchi, Tomoyuki	Osaka University, Japan
SaiToh, Akira	Kinki University, Japan
Sato, Jun	Ochanomizu University, Japan
Shiba, Noburo	Osaka University, Japan
Sugawa, Seiji	Kyoto University, Japan
Tai, Tasheng	Kinki University, Japan
Tamura, Ryo	University of Tokyo, Japan
Tanaka, Shu	University of Tokyo, Japan
Tim, Byrnes	NII, Japan
Tomita, Hiroyuki	Kinki University, Japan

ORGANIZING COMMITTEE

Mikio Nakahara   Kinki University (Chair)
Tasheng Tai   Kinki University
Shu Tanaka   University of Tokyo

CONTENTS

Preface   v

Symposium   vii

List of Participants   xi

Organizing Committee   xiii

Bosons in an Optical Lattice with a Synthetic Magnetic Field   3
    K. Kasamatsu

Quantum Simulation Using Exciton-Polaritons and their Applications Toward Accelerated Optimization Problem Search   39
    T. Byrnes, K. Yan, K. Kusudo, M. Fraser and Y. Yamamoto

Quantum Simulation Using Ultracold Atoms in Optical Lattices   61
    S. Sugawa, S. Taie, R. Yamazaki and Y. Takahashi

Universality of Integrable Model: Baxter's T-Q Equation, $SU(N)/SU(2)^{N-3}$ Correspondence and Ω-Deformed Seiberg-Witten Prepotential   83
    T.-S. Tai

Exact Analysis of Correlation Functions of the XXZ Chain   103
    T. Deguchi, K. Motegi and J. Sato

Classical Analogue of Weak Value in Stochastic Process   123
    H. Tomita

Scaling of Entanglement Entropy and Hyperbolic Geometry   143
    H. Matsueda

From Classical Neural Networks to Quantum Neural Networks   169
   B. Tirozzi

Analysis of Quantum Monte Carlo Dynamics in Infinite-range
Ising Spin Systems: Theory and Its Possible Applications   191
   J. Inoue

A Method to Control Order of Phase Transition: Invisible
States in Discrete Spin Models   217
   R. Tamura, S. Tanaka and N. Kawashima

Quantum Annealing and Quantum Fluctuation Effect in
Frustrated Ising Systems   241
   S. Tanaka and R. Tamura

BOSONS IN AN OPTICAL LATTICE
WITH A SYNTHETIC MAGNETIC FIELD

KENICHI KASAMATSU*

*Department of Physics, Kinki University,
Higashi Osaka City, Osaka 577-8502, Japan
* E-mail: kenichi@phys.kindai.ac.jp*

We briefly review the theoretical formulation of bosons in an optical lattice subject to a effective magnetic field. Starting from the Bose-Hubbard Hamiltonian, we can reduce the problem into the classical frustrated system in the two limiting cases, i.e., the Josephson junction regime and the hard-core limit at finite temperatures. The former can be described by the uniformly frustrated XY model, which is accessible in usual BEC experiments in a deep optical lattice. The latter introduces the extended XY model with the fluctuation of the amplitude of pseudospins, referred to as the gauged CP^1 model. The common features of the two systems are that the thermodynamic properties are sensitive to the magnetic flux piecing the plaquette. We show the results of the Monte Carlo simulations for the nontrivial latter case. Despite the presence of the particle number fluctuation, the thermodynamic properties are qualitatively similar to those of the frustrated XY model, where only the phase is a dynamical variable.

Keywords: Bosons; Optical Lattice; Synthetic magnetic field; Frustrated XY model; Phase transitions.

1. Introduction

Ultracold bosonic atoms in a laser-induced optical lattice (OL) give an ideal testing ground to study many-body effects associated with the model Hamiltonians in condensed matter systems.[1,2] The striking advantage is that the microscopic parameters of the OL, such as depth, periodicity, and symmetry can be precisely controlled. In the tight binding limit, a theoretical description of this system is given by the Bose-Hubbard model,[5] which describes, for example, the superfluid-Mott insulator transition as a representative of the strongly-correlated effect.[3,4] Despite the fact that the system is fully governed by the quantum nature, one can regard this system as a simulator of classical hamiltonians if the number of bosons at each lattice site is relatively large. Then, many Bose-Einstein condensates (BECs) are separated by potential barriers along the lattice direction, forming a bosonic Josephson junction array (JJA).[6,7] A two-dimensional (2D) JJA is a well-controlled system for the study of nontrivial phase transitions as well as macroscopic quantum phase coherence,[8] being governed by the classical XY model. It has suggested that BECs confined by a 2D OL can mimic the physics of 2D JJAs.[9] Recently, thermally-activated vortex formation, related with the Berezinskii-Kosterlitz-Thouless (BKT) mechanism, in such a 2D bosonic JJA was observed.[10]

When a magnetic field is inserted to the system, we can expect versatile cold-atom simulators of classical or quantum models, which can demonstrate various effects such as quantum Hall effects.[11,12] However, since charge of the cold atoms is neutral, their orbital motion does not respond to a real magnetic field. Analogous effects of a magnetic field can be gained by applying a rotation. Two experiments were reported, making use of a rotating OL to study quantized vortices in gaseous BECs.[13,14] Moreover, a staggered "effective" magnetic field was demonstrated in the recent experiment by using Raman-assisted tunneling in an optical superlattice for cold atoms.[15] These systems are described by the modified Bose-Hubbard model under an effective magnetic field, which exhibits very interesting physics,[16–30] far beyond the usual Bose-Hubbard model. These properties are inherited from the remarkable structure of the energy spectrum for noninteracting problems, i.e., a single particle moving on a tight-binding lattice in the presence of a uniform magnetic field. Here, the energy spectrum depends sensitively on the frustration parameter f, a magnetic flux per plaquette of the lattice, and exhibits a fractal structure known as the "Hofstadter butterfly".[31] For a rational value $f = p/q$ (with the integers p and q), there are q bands, and each state is q-fold degenerate.[28,32,33] In the strongly interacting regime, where both the particle number per site and the magnetic flux per plaquette are of order unity, it is theoretically predicted that there exist strongly correlated phases representative of the continuum quantum Hall states.[18,23,25]

In this paper, we focus on the classical regime of the OL system with an effective magnetic field. In the weakly-interacting condensed phase, the JJA under a magnetic field provide a close analog of that system, and implementations using cold atoms have been proposed.[34,35] The uniformly frustrated XY model (UFXYM) has been used to study such a JJA, being written by

$$H = -J \sum_{\langle i,j \rangle} \cos(\theta_i - \theta_j + A_{ij}). \tag{1}$$

Here, $J > 0$ denotes the coupling constant, θ_i the phase of the superconducting node at a site i, and $\langle i, j \rangle$ nearest neighbors. The bond variables A_{ij} satisfy the constraint $\sum A_{ij} = 2\pi f$ with the frustration parameter f, where the summation is taken over the perimeter of a plaquette. In the case of the superconductor, f is the magnetic flux piecing the plaquette in units of the flux quantum. The competition of two length scales — the mean separation of magnetic fluxes (vortices) induced by the frustration and the period of underlying lattice — yields a rich variety of the ground state structures and phase transitions, which depend on the rational or irrational values of the frustration parameter f.[36-41]

Starting from the Bose-Hubbard model, we can reduce the problem into the classical frustrated system in the two limiting cases, i.e., the Josephson junction regime and the hard-core limit at finite temperatures. The former can be described by the uniformly frustrated XY model, which is accessible in usual BEC experiments in a deep optical lattice. Hard core bosons are available for cold atom systems with low densities and very strong repulsive interactions, e.g., tuned by the Feshbach resonance.[42] In the hard-core limit, the Bose-Hubbard model can be mapped to the quantum spin model and described by the CP^1 (complex projective) operators, which are useful to construct the path-integral formulation. In the high-temperature limit, the quantum Hamiltonian reduces to the classical XY model with the fluctuation of the pseudospin amplitude, frustrated by the gauge field, which is referred to as the gauged CP^1 model below. This reduction provides a practical platform to explore the finite-temperature phase diagram of this system and to discuss the detailed critical properties of the phase transitions. The common features of the two systems are that the thermodynamic properties are sensitive to the magnetic flux piecing the plaquette. We show the results of the Monte Carlo simulations for the nontrivial latter case. Despite the presence of the particle number fluctuation, the thermodynamic properties are qualitatively similar to those of the frustrated XY model, where only the phase is a dynamical variable.

The paper is organized as follows. Section 2 introduces the basic formulation of our problem and derive the UFXYM for the 2D bosonic JJA and the CP^1 model for the hard-core bosons, starting from the Bose-Hubbard model. The ground-state properties of the gauged CP^1 model are discussed in Sec. 3. In Sec. 4, based on the Monte-Carlo simulations, we study the finite-temperature phase structures of the gauged CP^1 model for the averaged site occupation of hard-core bosons being half-filled. Section 5 is devoted to summary of this work.

2. Formulation

2.1. *Bose-Hubbard model*

The many-body Hamiltonian of the boson system subject to an "effective magnetic field" is given by

$$\hat{H} = \int d\mathbf{r}\, \hat{\psi}^\dagger \left[\frac{(-i\hbar\nabla - q^*\mathbf{A})^2}{2m} + V_{\text{ex}} + \frac{g}{2}\hat{\psi}^\dagger\hat{\psi} - \mu \right] \hat{\psi}, \quad (2)$$

where m is the atomic mass and $g = 4\pi\hbar^2 a/m$ the coupling constant with the s-wave scattering length a. The boson field operator $\hat{\psi}$ obeys the commutation relation $[\hat{\psi}(\mathbf{r}), \hat{\psi}^\dagger(\mathbf{r}')] = \delta(\mathbf{r} - \mathbf{r}')$. The conservation of the total particle number is ensured by the chemical potential μ. In this paper, we assume that the external potential consists of two parts $V_{\text{ex}} = V_{\text{ho}} + V_{\text{OL}}$ with an axisymmetric harmonic potential

$$V_{\text{ho}} = \frac{1}{2}m\omega_\perp^2 r^2 + \frac{1}{2}m\omega_z^2 z^2 \quad (3)$$

and a 2D OL

$$V_{\text{OL}} = V_0[\sin^2(kx) + \sin^2(ky)] \quad (4)$$

with the square lattice geometry and the spatial periodicity $d = \pi/k$. The minima of the OL are located at the points $id \equiv (i_x, i_y)d$ with the integers i_x and i_y. If we consider $\omega_\perp \ll \omega_z$, the confining potential becomes a pancake shape, realizing a quasi-2D regime.

Since the charge of the cold atoms is neutral, q^* represents the fictitious charge and $\mathbf{A}$ is the effective vector potential. Typical methods to create an analog of the magnetic field are as follows. (i) Applying a rotation to the system is a natural way to realize this situation. This is because the neutral atoms in a rotating reference frame experience a Coriolis force of the same form as the Lorentz force on charged particles in a magnetic field. Then, the term $q^*\mathbf{A}$ is interpreted as $m\mathbf{\Omega} \times \mathbf{r}$. The centrifugal force also acts the atoms so that the frequency of the trapping potential is modified, e.g., when $\mathbf{\Omega} = \Omega\hat{\mathbf{z}}$, $\omega_\perp \to \omega_\perp - \Omega$. This radial expansion of the condensed cloud yields the difficulty to achieve an ultrafast rotating regime of bosons, where strongly-correlated quantum Hall phase would emerge.[12] (ii) There are several proposals to generate the artificial gauge fields for neutral atoms without[43–47] or with OL,[16–19] by using the properly designed laser field (see[11] for review). The first experiment on the "synthesize" of gauge fields was made in the NIST group. In the experiment by Lin et al.,[48] the effective vector potential is generated by coupling a pair of Raman laser beams into the magnetic sublevels of the $F = 1$ hyperfine level of the optically trapped BEC of ^{87}Rb atoms. Successively, the synthetic magnetic[49] and electric fields[50] were also produced from a spatial variation and time dependance of the effective vector potential, respectively. In addition, by using the similar scheme, BEC with spin-orbit coupling has also been realized by the same group.[51] Fu et al. reported the creation of an effective gauge potential for $F = 2$ ^{87}Rb BEC.[52]

In the following, we will assume that the amplitude of the OL is large enough to create many separated wells giving rise to a 2D array of condensates. Still, because of quantum tunneling, the overlap between the wave functions of two consecutive wells can be sufficient to ensure full coherence. If the energy due to interaction and effective magnetic field (rotation) is small compared to the energy separation between the lowest and first excited band, the particles are confined to the lowest Wannier orbitals. Following the fact that the first term of Eq. (2) is analogous to the Hamiltonian of an electron in a magnetic field, we take the Wannier basis as

$$\hat{\psi}(\mathbf{r}) = \sum_i \hat{a}_i w_i(\mathbf{r}) \exp\left[\frac{iq^*}{\hbar} \int_{\mathbf{r}_i}^{\mathbf{r}} \mathbf{A}(\mathbf{r}') \cdot d\mathbf{r}'\right], \quad (5)$$

where $\mathbf{A} = \mathbf{\Omega} \times \mathbf{r}$ is the analog of the magnetic vector potential, $w_i(\mathbf{r})$ the Wannier wave function localized at the ith well and $\hat{a}_i^{(\dagger)}$ the boson annihilation (creation) operator on the site i. Since the total particle number is given by

$$N = \int d\mathbf{r} \langle \psi^\dagger(\mathbf{r}) \psi(\mathbf{r}) \rangle = \sum_{i,j} \langle \hat{a}_i^\dagger \hat{a}_j \rangle \int d\mathbf{r} w_i^*(\mathbf{r}) w_j(\mathbf{r}), \quad (6)$$

$w_i(\mathbf{r})$ should satisfy the normalization condition $\int dr w_i^*(\mathbf{r}) w_j(\mathbf{r}) = \delta_{ij}$, thereby $N = \sum_i \langle \hat{a}_i^\dagger \hat{a}_i \rangle \equiv \sum_i \langle \hat{\rho}_i \rangle \equiv \sum_i \rho_i$, where $\hat{\rho}_i = \hat{a}_i^\dagger \hat{a}_i$ is the number operator at i-site.

We consider a system of N-bosons put on the sites of a 2D square lattice with the size $L \times L$. Substituting Eq. (5) in the Hamiltonian Eq. (2) leads to the 2D Bose-Hubbard model subject to a uniform abelian gauge potential[16–30]

$$\hat{H} = -\sum_{\langle i,j \rangle} \frac{t_{ij}}{2} \left[\hat{a}_i^\dagger \hat{a}_j e^{iA_{ij}} + \text{h.c.} \right] + \sum_i \frac{U_i}{2} \hat{\rho}_i(\hat{\rho}_i - 1) + \sum_i (\epsilon_i - \mu) \hat{\rho}_i. \quad (7)$$

Here, $\langle i, j \rangle$ implies a pair of nearest-neighbor sites and

$$t_{ij} = -\int d\mathbf{r} w_i^*(\mathbf{r}) \left(-\frac{\hbar^2}{2m}\nabla^2 + V_{\mathrm{OL}}\right) w_j(\mathbf{r}), \tag{8}$$

$$\epsilon_i = \int d\mathbf{r} w_i^*(\mathbf{r}) \left(-\frac{\hbar^2}{2m}\nabla^2 + V_{\mathrm{ex}} - \mu\right) w_i(\mathbf{r}), \tag{9}$$

$$U_i = g \int d\mathbf{r} |w_i(\mathbf{r})|^4. \tag{10}$$

represent the hopping matrix element, the energy offset of each lattice site, and the on-site energy, respectively. The Hamiltonian conserves the total number of bosons $\hat{N} = \sum_i \hat{\rho}_i$. The energy offset ϵ_i is caused by the additional trap potentials such as a harmonic one. The results for the case of $\epsilon_i = 0$ can be applied within the local density approximation to realistic experimental systems which have an additional trapping potential.

The field A_{ij} describes the imposed gauge potential, defined by $A_{ij} = \int_{\mathbf{r}_j}^{\mathbf{r}_i} \mathbf{A} \cdot d\mathbf{r}$. All of the physics of the system governed by the Hamiltonian (7) is gauge-invariant. Hence, its properties depend only on the magnetic fluxes of magnitude B through plaquettes

$$\Phi = \int_{\mathrm{plaq}} d\mathbf{S} \cdot \mathbf{B} = \sum_{i,j \in \alpha} A_{ij} = Bd^2 \equiv f\Phi_0 \tag{11}$$

where α labels the plaquette, and the sum represents the directed sum of the gauge fields around that plaquette (the discrete version of the line integral). Φ_0 represents a flux of quantum and $f = \Phi/\Phi_0$ the number of the magnetic flux per plaquette. In the case of a condensate with rigid-body rotation, the number of vortices per unit area is $2\Omega/\kappa$ with the quantum circulation $\kappa = h/m$. Then, f is given by the average number of vortices per unit cell of the OL: $f = 2\Omega d^2/\kappa$. In the following, we use the vector potential in the symmetric gauge and its form can be written as

$$\mathbf{A} = (A_x, A_y, A_z) = \left(-\frac{B}{2}y, \frac{B}{2}x, 0\right). \tag{12}$$

This corresponds to a uniform magnetic field $\mathbf{B} = (0, 0, B)$ in the direction perpendicular to the lattice plane. The discrete vector potential is then written as

$$A_{ij} = \begin{cases} -\pi f i_y & \text{for } i = (i_x, i_y), \ j = (i_x + 1, i_y) \\ \pi f i_x & \text{for } i = (i_x, i_y), \ j = (i_x, i_y + 1). \end{cases} \tag{13}$$

Note that the gauge field A_{ij} is constant in the present model. This should be distinguished with the related "gauge models"[53,54] in which dynamical U(1) gauge field is involved.

2.2. Frustrated XY model

The formulation of BECs in a deep 2D OL can be mapped into the XY model, where the amplitude of the condensate wave function is frozen at each site but its phase is still a relevant dynamical variable.[9] If the number of atoms per site is large ($N_i \gg 1$), the operator can be expressed in terms of its amplitude and phase, the amplitude being subsequently approximated by the c-number as $\hat{a}_i \simeq \sqrt{N_i} e^{i\theta_i}$. Then, the Bose-Hubbard model Eq. (7) reduces to

$$\hat{H} = -\sum_{\langle i,j \rangle} J_{ij} \cos(\theta_i - \theta_j + A_{ij}) - \sum_i \frac{U_i}{2} \frac{\partial^2}{\partial \theta_i^2}$$
$$-i\sum_i (E_i + U_i N_i) \frac{\partial}{\partial \theta_i} + \sum_i \left(E_i N_i + \frac{U_i}{2} N_i^2 \right), \qquad (14)$$

where we have used the phase representation $\hat{N}_i = N_i + \delta \hat{N}_i = N_i - i\partial/\partial \theta_i$, $\hat{\theta}_i = \theta_i$ and the notation $J_{ij} = 2t_{ij}\sqrt{N_i N_j}$. This reduction is valid when $J_{ij}/N_i^2 \ll U_i$,[9] which is well satisfied because of $N_i \gg 1$.

The first term of Eq. (14) describes the UFXYM with spatially inhomogeneous coupling constant J_{ij}. As discussed below, the equilibrium condition for the density is determined by minimizing the last term of Eq. (14). Then, $E_i + U_i N_i = 0$ and the last two terms may not be important for the dynamics of the phase θ_i. Furthermore, if $J_{ij} \gg U$ is satisfied, the system can be mapped into the UFXYM. One can see in the Appendix A that this regime can be accessible in the typical BEC experiment in an optical lattice. Then, the ground state and the thermodynamic properties are expected to be qualitatively similar to the UFXYM with uniform coupling constant J.[34,35] The additional potential (energy offset) makes the system finite size, providing a new concept of the so-called "trap-size scaling".[55]

It is known that the UFXYM exhibits very rich properties of the phase transition.[56–70] Here, "frustration" refers to the fact that, with $f \neq 0$ for any plaquette, the angles θ_i around this plaquette cannot be chosen to maximally satisfy the XY exchange couplings. For the fully frustrated XY model (FFXYM) with $f = 1/2$, there would be double phase transitions, one is associated with the BKT transition due to the global U(1) symmetry and the other associated with Ising-model-like transition due to the Z_2 chiral symmetry of the ground state. A central issue in the studies of the FFXYM has been to clarify how these two distinct types of orderings take place.[56] One possibility is that, even at the temperature where the Z_2 chirality establishes a long-range order at $T < T_c$, the U(1) phase (XY-spin) may remain disordered due to thermally excited, unbound vortices. Then, the orderings of the two variables take place at two separate temperatures such that $T_c > T_{\text{BKT}}$. The other possibility is that both orderings of the chirality and the phase take place at the same temperature, and the resulting single phase transition is neither of the conventional Ising type nor the BKT type but follows a new universality class. Many studies have discussed this phase transition, but do not provide yet conclusive results. The phase transitions for other values of f, e.g. $f = 1/3$, $2/5$, etc., were also discussed in Refs.[67–70] ; their nature is dominated by the properties of the domain walls, following the Ising-like transition for $f = 1/3$ and the first-order transition for $f = 2/5$.

2.3. *Hamiltonian for hard-core bosons in an effective magnetic field*

In contrast to the previous subsection, we here take a hard-core limit ($U \to \infty$), where the allowed physical states at i-site are eigenstates of $\hat{\rho}_i$ with eigenvalue 0 or 1 and their superpositions. States with higher particle number at the same site such as double occupancy are excluded. We introduce a destruction operator of the hard-core boson as $\hat{\phi}_i$, which satisfies the following mixed canonical-(anti)commutation relations

$$\left[\hat{\phi}_i, \hat{\phi}_j\right] = 0, \quad \left[\hat{\phi}_i, \hat{\phi}_j^\dagger\right] = 0 \quad \text{for } i \neq j, \tag{15}$$

and on the same site,

$$\left\{\hat{\phi}_i, \hat{\phi}_i\right\} = 0, \quad \left\{\hat{\phi}_i, \hat{\phi}_i^\dagger\right\} = 1. \tag{16}$$

Thus, the number operator is rewritten as $\hat{\rho}_i = \hat{\phi}_i^\dagger \hat{\phi}_i$ and its eigenvalue ρ_i is assured to be 0 or 1. Then, the Hamiltonian $\hat{H}$ is written as

$$\hat{H}_{\text{hc}} = -\frac{t}{2} \sum_{\langle i,j \rangle} \left(\hat{\phi}_i^\dagger \hat{\phi}_j e^{iA_{ij}} + \text{h.c.}\right) - \mu \sum_i \hat{\phi}_i^\dagger \hat{\phi}_i, \tag{17}$$

where μ determines the mean density ρ of hard-core bosons per site,

$$\rho \equiv \langle \bar{\rho} \rangle, \quad \bar{\rho} \equiv \frac{1}{L^2} \sum_i \hat{\phi}_i^\dagger \hat{\phi}_i, \tag{18}$$

within the range $0 \leq \rho \leq 1$. We also omitted the additional confining potential in the following.

In the hard-core limit, the Bose-Hubbard model becomes equivalent to a spin-1/2 quantum magnet, because the following relations exist between

$\hat{\phi}_i$ and the $s = 1/2$ SU(2) spin operator $\hat{s}_i^{x,y,z}$,[53]

$$\hat{s}_i^z = \hat{\phi}_i^\dagger \hat{\phi}_i - \frac{1}{2}, \quad \hat{s}_i^+ \equiv \hat{s}_i^x + i\hat{s}_i^y = \hat{\phi}_i^\dagger, \quad \hat{s}_i^- = \hat{\phi}_i,$$

$$(\hat{s}_i^x)^2 + (\hat{s}_i^y)^2 + (\hat{s}_i^z)^2 = \frac{3}{4}. \tag{19}$$

The Hamiltonian (17) then becomes

$$\hat{H}_{\text{hc}} = -\frac{t}{2} \sum_{\langle i,j \rangle} \left(\hat{s}_i^+ \hat{s}_j^- e^{iA_{ij}} + \hat{s}_j^+ \hat{s}_i^- e^{-iA_{ij}} \right) - \mu \sum_i \hat{s}_i^z. \tag{20}$$

Here, the conservation of total particle number is interpreted as the constant magnetization $\hat{S}^z = \sum_i \hat{s}_i^z$. Under Eq.(19), the eigenstates have the correspondence $|\uparrow_i\rangle = |\rho_i = 1\rangle$ and $|\downarrow_i\rangle = |\rho_i = 0\rangle$. This Hamiltonian describes a quantum spin-1/2 magnet, experiencing XY nearest neighbor spin exchange interactions. These exchange interactions are frustrated due to the gauge field A_{ij}.

2.3.1. CP^1 variable and path-integral representation

To study the thermodynamic properties of this system, we evaluate the partition function Z of the grand canonical ensemble,

$$Z = \text{Tr} e^{-\beta \hat{H}_{\text{hc}}}, \quad \beta = \frac{1}{k_B T}. \tag{21}$$

We express Z by a path integral which is useful for numerical calculations. For this purpose, it is convenient to introduce a CP^1 variable $w_i = (w_{1i}, w_{2i}) \in C$ which satisfies the CP^1 constraint,

$$|w_{1i}|^2 + |w_{2i}|^2 = 1. \tag{22}$$

An associated pseudo-coherent state $|w_i\rangle$ is defined by

$$|w_i\rangle \equiv w_{1i} |\uparrow_i\rangle + w_{2i} |\downarrow_i\rangle, \tag{23}$$

where $|w_i\rangle$ is normalized as $\langle w_i | w_i \rangle = 1$ due to Eq. (22), and we have generally $\langle w_i | w_i' \rangle = w_{i1}^* w_{i1}' + w_{i2}^* w_{i2}' = w_i^* w_i'$. Let us define the integration measure

$$\int [d^2 w_i] \equiv 2 \int_C d^2 w_{1i} \int_C d^2 w_{2i} \delta(\langle w_i | w_i \rangle - 1) \tag{24}$$

which satisfies

$$\int [d^2 w_i] 1 = 2, \quad \int [d^2 w_i] w_{ia}^* w_{ib} = 2 \times \frac{1}{2} \delta_{ab} = \delta_{ab}. \tag{25}$$

Then the completeness is expressed as

$$\int [d^2 w_i]|w_i\rangle\langle w_i| = |\uparrow_i\rangle\langle\uparrow_i| + |\downarrow_i\rangle\langle\downarrow_i| = 1. \tag{26}$$

The Hamiltonian can be represented by the CP^1 operators $\hat{w}_{1i}$ and $\hat{w}_{2i}$ which satisfy the bosonic commutation relation

$$[\hat{w}_{ai}, \hat{w}_{bj}] = 0, \quad [\hat{w}_{ai}, \hat{w}_{bj}^\dagger] = \delta_{ab}\delta_{ij}, \quad a,b = 1,2, \tag{27}$$

and their physical states are restricted as $\sum_a \hat{w}_{ai}^\dagger \hat{w}_{ai}|\text{phys}\rangle = |\text{phys}\rangle$. Here, the correspondence to the spin operator is given as

$$|\uparrow_i\rangle = \hat{w}_{1i}^\dagger|\text{vac}\rangle, \quad |\downarrow_i\rangle = \hat{w}_{2i}^\dagger|\text{vac}\rangle, \tag{28}$$

and

$$\hat{s}_i^{x,y,z} = \frac{1}{2}(\hat{w}_{1i}^\dagger, \hat{w}_{2i}^\dagger)\sigma^{x,y,z}(\hat{w}_{1i}, \hat{w}_{2i})^t,$$

$$\hat{s}_i^z = \hat{w}_{1i}^\dagger \hat{w}_{1i} - \frac{1}{2}, \quad \hat{s}_i^+ = \hat{w}_{1i}^\dagger \hat{w}_{2i}, \quad \hat{s}_i^- = \hat{w}_{2i}^\dagger \hat{w}_{1i}, \tag{29}$$

where $\sigma^{x,y,z}$ are Pauli matrices. Equations (29) and (19) imply the relation $\hat{\phi}_i = \hat{w}_{2i}^\dagger \hat{w}_{1i}$, etc. We also note the following relation,

$$\langle w_i'|\hat{w}_{ia}^\dagger \hat{w}_{ib}|w_i\rangle = (w_{ia}')^* w_{ib}. \tag{30}$$

Using these relations and following the standard procedure, we can write the partition function Z in the path-integral form with the imaginary time $\tau \in [0,\beta]$ as

$$Z = \prod_{i,\tau} \int [d^2 w_i(\tau)] e^{\int_0^\beta d\tau A(\tau)}, \tag{31}$$

where

$$A(\tau) = -\sum_{i,a} w_{ai}^*(\tau)\dot{w}_{ai}(\tau) + \frac{t}{2}\sum_{\langle i,j\rangle}\left(w_{1i}^*(\tau)w_{2i}(\tau)w_{2j}^*(\tau)w_{1j}(\tau)e^{iA_{ij}}\right.$$

$$\left. + \text{c.c.}\right) + \mu\sum_i w_{1i}^*(\tau)w_{1i}(\tau). \tag{32}$$

Here, the CP^1 operators have been replaced to the complex numbers by employing the path-integral formulation.

To proceed further, we make one simplification by considering the finite-T region, such that the τ-dependence of $w_{ai}(\tau)$ in the path integral can be ignored keeping only the zero modes as $w_{ai}(\tau) \to w_{ai}$.[54] This corresponds to neglecting the quantum fluctuations. As a result, the problem is reduced to the classical one. Under the relations (19) and (29), one can finally obtain

$$Z = \prod_i \int [d^2 w_i] e^{-\beta H_{\text{CP}^1}(w)}, \tag{33}$$

where $H_{\text{CP}^1}(w)$ is the classical version of Eq. (17) expressed in terms of w_i:

$$H_{\text{CP}^1}(w) = -\frac{t}{2} \sum_{\langle i,j \rangle} \left(\phi_i^* \phi_j e^{iA_{ij}} + \text{c.c.} \right) - \mu \sum_i w_{1i}^* w_{1i},$$

$$\phi_i \equiv w_{2i}^* w_{1i}. \tag{34}$$

This allows us to study the finite-T phase structure, which summarizes the essential properties of the system. Besides, the finite-T phase diagram gives a very useful insight into the phase structure at $T = 0$; if some ordered states are found at finite T, we can naturally expect that they persist down to $T = 0$. Equation (34) is referred to as the "gauged CP^1 model" in the following. We can evaluate the partition function of Eq. (33) to understand the thermodynamic properties of the system, using the standard Monte Carlo simulations. From the symmetry properties summarized in Appendix C, we can confine ourselves to $0 \leq f \leq 1/2$ and $1/2 \leq \rho \leq 1$ in the following argument.

To make clear the relation with the XY model, we rewrite the CP^1 variables as

$$w_i = \begin{pmatrix} w_{1i} \\ w_{2i} \end{pmatrix} = \begin{pmatrix} \cos(\psi_i/2) e^{i\lambda_{1i}} \\ \sin(\psi_i/2) e^{i\lambda_{2i}} \end{pmatrix}. \tag{35}$$

Here, the angle variables have the ranges $0 \leq \psi_i \leq \pi$, $0 \leq \lambda_{1i,2i} \leq 2\pi$. The Hamiltonian Eq. (34) can be written as

$$H_{\text{CP}^1} = -\frac{t}{4} \sum_{\langle i,j \rangle} \sin \psi_i \sin \psi_j \cos(\theta_i - \theta_j + A_{ij}) - \frac{\mu}{2} \sum_i \cos \psi_i \tag{36}$$

with $\theta_i = \lambda_{2i} - \lambda_{1i}$ and the integration measure $[d^2 w_i] = (4\pi^2)^{-1} \sin \psi_i d\psi_i d\lambda_{1i} d\lambda_{2i}$. If we restrict the configuration space with fixed $\psi_i = \pi/2$, the ground state has (uniform) density $\rho = 1/2$, so that $s_i^z = 0$ and all the spins lie in the xy-plane. Then, the Hamiltonian reduces to the FXYM Eq. (1). The CP^1 model has a site-dependent factor $\sin \psi_i \sin \psi_j$ associated with the variation of the particle number at each site. Hence, the CP^1 model includes the particle number fluctuation and goes beyond the XY model based on the phase fluctuation only.

In the following, we compare the thermodynamic properties of our gauged CP^1 model Eq. (36) and the FXYM Eq. (1). The latter has been extensively studied for decades.[56–66] To this end, we have to put the two models in the same energy measure by establishing a relation between J and t. Let us try to replace the density at site i in Eq. (36) to the average value ρ, i.e., $\sin\psi_i \sin\psi_j \to \sin^2\psi$ with

$$\sin^2\psi = 4\cos^2\frac{\psi}{2}\left(1 - \cos^2\frac{\psi}{2}\right) = 4\rho(1-\rho), \qquad (37)$$

where we have used $\rho \simeq \langle \phi^*\phi \rangle = \langle w_1^* w_1 \rangle = \cos^2(\psi/2)$. This correspondence implies the relation $J = t\rho(1-\rho)$. We use this relation when the energies in the two models (1) and (36) are compared. Note that this relation reflects the particle-hole symmetry and is *different from* the naive replacement $J = t\rho$.

3. Ground state

We discuss here the ground-state properties of the CP^1 model. It is expected that the ground state exhibits similar behaviors to the UFXYM, where the magnetic flux forms the typical pattern with the $q \times q$ unit cell structures for $f = p/q$; the checkerboard pattern emerges for $f = 1/2$ and the staircase pattern for $1/3 \leq f \leq 1/2$.[36–38] Since the CP^1 model has an additional degree of freedom associated with the density fluctuation, it is expected that there are some differences from the results of the FXYM.

We calculate numerically the ground state of the gauged CP^1 model for fixed ρ by the simulated annealing method. Two examples for $\rho = 0.5$ and 0.95 are shown in Fig. 1, where the former corresponds to $\mu = 0$ (see Appendix C) while the latter is obtained by adjusting a proper value of μ for a given f. Figures 1 (a) and (b) represent the ground-state energy $E_{\min}$ as a function of f. We also plot the energy for the FXYM for comparison. The shape of the energy curve is non-monotonic behavior with respect to f, following the bottom of the energy spectrum of the Hofstadter butterfly.[38]

For $\rho = 0.5$ the ground-state energy coincides completely with that of the XY model. This is clear because ψ_i is freezed to $\pi/2$ at $\rho = 0.5$ and the particle number does not fluctuate. Because $s_i^z = 0$ there, the ground-state properties are not affected at all even for $f \neq 0$. As ρ deviates from 0.5, the ground-state energy of the CP^1 model becomes slightly lower than that of the XY model, except for $f = 0.5$, as shown in Fig. 1(b).

Figures 1(c) and (d) show the distribution of the mean density $\rho_{\bar{i}}$ and the vorticity $m_{\bar{i}}$ at the site of the dual lattice $\bar{i}$, defined by

$$\rho_{\bar{i}} = \frac{1}{4}\sum_{i \in \alpha} \rho_i \quad m_{\bar{i}} = \frac{1}{2\pi}\sum_{i,j \in \alpha}(\theta_i - \theta_j + A_{ij}), \qquad (38)$$

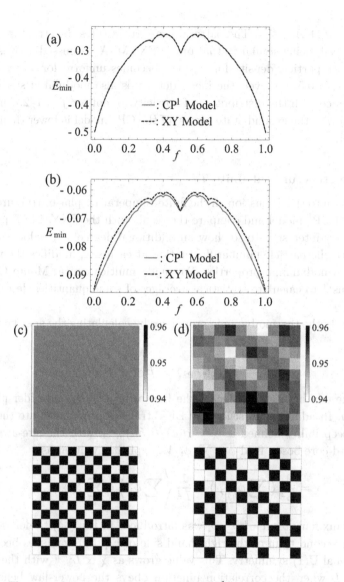

Fig. 1. The ground-state energy per site of the CP^1 model Eq. (36) as a function of the magnetic flux f for the averaged density (a) $\rho = 0.5$ and (b) 0.95, obtained through the simulated annealing. The solid curve denotes the energy of the CP^1 model without the chemical-potential term, while the dashed curve denotes that of the XY model with setting $J = t\rho(1-\rho)$ for comparison. The two curves overlap completely in (a); see the text for the reason. (c) and (d) show the distribution over the lattice ($L = 12$ and 10 for $f = 1/2$ and $2/5$, respectively) of the mean density $\langle \rho_{\bar{i}} \rangle$ (the upper panels) and the vorticity $m_{\bar{i}}$ (the lower panels) at each plaquette $\bar{i}$ (the site of the dual lattice) for $\rho = 0.95$; (c) $f = 1/2$ and (d) $f = 2/5$.

where $|\theta_i - \theta_j + A_{ij}| \leq \pi$. The pattern of m_α constitutes the structure of a $q \times q$ unit cell, being similar to that of the FXYM. As discussed above, the ground-state particle density for $\rho = 0.5$ becomes uniform for any values of f. For $\rho \neq 0.5$, however, the mean density is also modulated spatially in accordance with the distribution of vorticity except for $f = 1/2$. This is the reason why the ground-state energy of the CP^1 model is lower than the XY model.

4. Phase structures at finite T

We next turn to the discussion on the finite-temperature phase structures of the gauged CP^1 model and compare the result with the FXYM of Eq. (1). Our primary interest is to see how an additional degree of freedom, associated with the particle number fluctuation at each site, modifies the well-known thermodynamic properties. We made multicanonical Monte Carlo simulations[71] to calculate statistical averages of some quantities described below.

We study the thermodynamic properties by calculating the specific heat defined by

$$C = \frac{1}{L^2} \left(\langle H_{\text{CP}^1}^2 \rangle - \langle H_{\text{CP}^1} \rangle^2 \right). \tag{39}$$

This value can be useful to locate the first-order and second-order phase transition. In addition, to study the BKT transition, we calculate the in-plane susceptibility defined as $\chi = \partial \langle \bar{\phi} \rangle / \partial(\beta h)|_{h \to 0}$ with the site-average of the hard-core boson field $\bar{\phi} = \sum_i \phi_i / V$, explicitly written as

$$\chi = \frac{1}{L^2} \left\langle \sum_{i,j} \phi_i^* \phi_j \right\rangle - \frac{1}{L^2} \left\langle \sum_i \phi_i^* \right\rangle \left\langle \sum_i \phi_i \right\rangle. \tag{40}$$

Here, an auxiliary term $-h \sum_i \phi_i$ was introduced in Eq. (34) to derive Eq. (40). The second term of the right-hand side of Eq. (40) vanishes because of the global U(1) symmetry. This value grows as $\chi \propto L^{2-\eta}$ with the system size L when the correlation function obeys the power-law behavior $\langle \phi_i^* \phi_j \rangle \propto r^{-\eta}$ (with $\eta \leq 2$) due to the presence of the quasi-long range order. On the other hand, it remains finite for $L \to \infty$ when the correlation decays exponentially as $\langle \phi_i^* \phi_j \rangle \propto e^{-mr}$. The value η is an exponent for the in-plane correlations below the BKT transition temperature T_{BKT}, being dependent on the temperature. We can calculate the exponent, as a function of temperature, by fitting the susceptibility for several system sizes L to the above expression for each temperature. For the conventional

XY model the critical temperature T_{BKT} can be estimated when χ grows as $L^{7/4}$, i.e., $\eta = 1/4$

Furthermore, we study the helicity modulus which is directly connected to the superfluid density. The helicity modulus Υ is a measure of the resistance to an infinitesimal spin twist $\Delta\theta$ across the system along one coordinate. More precisely, it is defined through the change of the total free energy F with respect to an infinitesimal twist on the spin configuration along, say, the x-axis $\theta_i - \theta_{i'} \to \theta_i - \theta_{i'} + \delta\theta$, where i' is the nearest-neighbor site of the i-site along the x-direction and $\delta\theta = \Delta\theta/L$. One readily finds

$$\Delta F = \Upsilon(\delta\theta)^2 + \mathcal{O}((\delta\theta)^4), \qquad (41)$$

and Υ can be given as

$$\Upsilon = -\frac{1}{L^2}\left(\langle H_x \rangle + \beta\langle I_x^2 \rangle\right), \qquad (42)$$

where H_x is the x-bond part of the Hamiltonian at $\Delta\theta = 0$ and I_x is the total current in the x-direction. For the CP^1 model, they are written as

$$H_x = -\frac{t}{4}\sum_{\langle i,j \rangle_x} \sin\psi_i \sin\psi_j \cos(\theta_i - \theta_j + A_{ij}),$$

$$I_x = -\frac{t}{4}\sum_{\langle i,j \rangle_x} \sin\psi_i \sin\psi_j \sin(\theta_i - \theta_j + A_{ij}). \qquad (43)$$

According to the renormalization-group theory, the helicity modulus for the conventional XY model in an infinite system jumps from zero to the finite value $(2/\pi)k_B T_{\text{BKT}}$ at the critical temperature $T = T_{\text{BKT}}$.[72] Therefore, a rough estimate of the critical temperature at a finite system could be obtained simply by locating the intersection of Υ as a function of T and the straight line $\Upsilon = 2k_B T/\pi$.

In the FFXYM, the jump size has been suggested to be lager than the universal value.[56–59,66] As a more accurate method to estimate both T_{BKT} and the jump size in the helicity modulus, an useful finite-size-scaling expression at $T = T_{\text{BKT}}$ is known as[73]

$$\Upsilon(L, T = T_{\text{BKT}}) = \frac{2}{\pi}k_B T^*_{\text{BKT}}\left(1 + \frac{1}{2}\frac{1}{\ln L + c}\right) \qquad (44)$$

with T^*_{BKT} and c being fitting parameters. T^*_{BKT} is related to the jump size, being equal to T_{BKT} in the case of the usual XY model. By making a fit of the numerical data to Eq. (44) at various temperatures, one can estimate the transition temperature T_{BKT} as well as the critical exponent from the jump size as $\eta = T_{\text{BKT}}/4T^*_{\text{BKT}}$.[66,74]

4.1. Density fluctuation

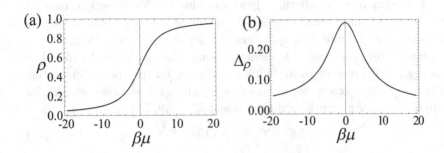

Fig. 2. The mean density ρ and the particle number fluctuation Δ_ρ as a function of $\beta\mu$ at the high-temperature or zero-hopping limit $\beta t \to 0$; the behavior is thus independent of f.

Before discussing the detailed thermodynamic properties, let us see the magnitude of the spatial fluctuation of the particle density, which is the important difference between the gauged CP^1 model and the FXYM. The particle density fluctuation is defined as

$$\Delta_\rho = \sqrt{\left\langle \frac{1}{L^2} \sum_i (\rho_i - \bar{\rho})^2 \right\rangle}, \qquad (45)$$

which is zero for the XY model. In the high-temperature limit $\beta t \to 0$, one can calculate exactly the partition function $Z = [(e^{\beta\mu} - 1)/\beta\mu]^V$, and thus

$$\rho = \frac{1 + (-1 + \beta\mu)e^{\beta\mu}}{\beta\mu(-1 + e^{\beta\mu})}, \qquad (46)$$

$$\Delta_\rho = \frac{-2 + (2 - 2\beta\mu + \beta^2\mu^2)e^{\beta\mu}}{\beta^2\mu^2(-1 + e^{\beta\mu})}, \qquad (47)$$

which are shown in Fig. 2. The mean density ρ is weakly dependent on βt for $\beta\mu \neq 0$. The density fluctuation Δ_ρ has a maximum at $\beta\mu = 0$, corresponding to half-filling $\rho = 1/2$, and decreases for $\beta\mu \to \pm\infty$. Since this Δ_ρ gives an upper limit of the expected particle number fluctuation, one can see that influence of the particle number fluctuation is less than 20% in a temperature range of our interest.

As seen in the ground state (Sec. 3), the density becomes uniform for $\mu = 0$. Thus, Δ_ρ should go to zero as $\beta t \to \infty$ for $\mu = 0$. On the other hands, for $\mu \neq 0$, it must remains finite because the ground state possesses

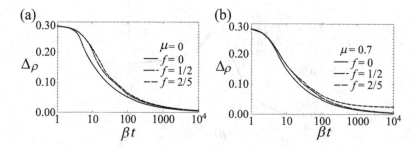

Fig. 3. The particle number fluctuation Δ_ρ as a function of βt for (a) $\mu = 0$ and (b) $\mu = 0.7$. The solid, dashed, and dotted curves correspond to $f = 0$, $1/2$, and $2/5$, respectively.

spatial density modulation. Figure 3 shows the βt-dependance of Δ_ρ for $\mu = 0$ and 0.7 for several values of f. For $\mu = 0$, Δ_ρ approaches to zero as $\beta t \to 0$ for any values of f. For $\mu = 0.7$, on the other hand, Δ_ρ approaches to zero as $\beta t \to 0$ only for $f = 1/2$, but remains finite for the other values of f. This behavior is consistent with the ground-state property shown in Fig. 1.

4.2. The finite temperature phase transition

In this paper, we consider the case of half-filling $\rho = 0.5$ by setting $\mu = 0$. Then, the density distribution is completely uniform in the ground state, as seen in the complete overlap of the ground-state energy in Fig. 1(a), and thus the CP^1 model reproduces the ground state of the XY model. However, one has to take into account the density fluctuation at finite temperatures. This additional degree of freedom makes the finite-temperature phase diagram and the nature of the phase transition nontrivial. Here we focus on the case $f = 0$, $1/2$ and $2/5$, each of which has been known to give rise to quite different phase transitions in the FXYM.

4.2.1. f=0

First, we consider the situation of a zero magnetic field $f = 0$, which is useful to confirm the accuracy of our numerical computation. Then, the model is equivalent to the XX0 (three-component XY) model studied in Ref.[79] The phase transition of this model has been found to be consistent with the BKT theory. The specific heat C has very small finite-size effects.

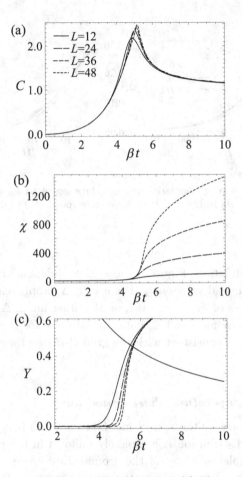

Fig. 4. Several thermodynamic quantities for $f=0$: (a) The specific heat C, (b) the in-plane susceptibility χ, and (c) the helicity modulus Υ for the CP^1 model with $\mu = 0$ and $f = 0$ as a function of βt. The system size for each curve is $L = 12, 24, 36, 48$. An additional curve $\Upsilon = 8/\pi\beta t$ in (c) indicates the universal jump of a BKT transition.

The in-plane susceptibility χ is a strongly increasing function of the system size L for low temperatures, while all the data fall on the same curve for high temperatures. These facts indicate the absence of second-order transition and the possibility of the BKT type transition involving the power-law decay of the correlation function. The transition temperature of the XX0 model was obtained as $T_{\text{BKT}} = 0.699J/k_{\text{B}}$,[79] which is lower than the usual (two-component) XY model $T_{\text{BKT}} = 0.898J/k_{\text{B}}$.[75] This is naturally understood due to the difference of the degree of freedom of these models.

Our numerical result is shown in Fig. 4. The size dependance of C is actually small. We estimate the BKT transition temperature T_{BKT} by two methods, one by using the data of χ and the other by using Υ. We plot $\chi/L^{7/4}$ for some values of L as a function of βt by assuming $\eta = 1/4$, in which the crossing point of the curves gives T_{BKT}. From Υ, we take the temperatures that correspond to the crossing points of $\Upsilon(T)$ and the line $\Upsilon = (2/\pi)k_\text{B}T_{\text{BKT}}$ related to the universal jump value for various values of L, interpolating them to $L \to \infty$. Both these methods give the same $T \approx 0.702 J/k_\text{B}$, which is consistent with the result of Ref.[79] and confirms the accuracy of our numerical computations.

4.2.2. *f=1/2*

We next consider the case with full frustration $f = 1/2$. If there is a continuous phase transition, various quantities should exhibit a singular behavior near the transition temperature T_c. Figure 5(a) represents the specific heat C as a function of βt, where C exhibits a peak structure as a function of βt and the height of the peak increases with increasing L. This suggests the occurrence of a second-order phase transition. Concurrently, χ and Υ grow from zero with increasing βt, which is a signature of the emergence of superfluid order. Hence, the qualitative feature of the phase transition is similar to the FFXYM.

It is important to clarify whether the nature of the phase transition is consistent with those observed in the analysis of the FFXYM. In the FFXYM, two separate phase transitions may occur, corresponding to the breaking of the Z_2 chirality and the U(1) symmetry.[56–66] It is still inconclusive whether the former obeys universality class of the Ising transition and the latter is subject to the BKT mechanism with non-universal jump of the helicity modulus.

First, we focus on the data of the specific heat to clarify the Z_2-related phase transition. To determine the critical temperature and the critical exponents, we use finite-size scaling analysis,[76] where the specific heat C is expressed as

$$C(L, \tilde{t}) = L^{\alpha/\nu} \tilde{\phi}(L^{1/\nu} \tilde{t}), \tag{48}$$

where $\tilde{t} = (T - T_c)/T_c$ is a reduced temperature, α the standard exponent for the specific heat, and ν the exponent for the divergence of the correlation length. Since the scaling functions $\tilde{\phi}$ should depend on a single variable, we can make all the data for each system size L fall on the same curve by

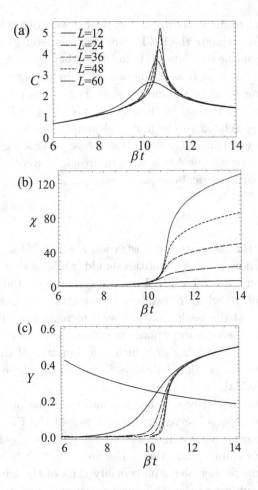

Fig. 5. Several thermodynamic quantities for $f=1/2$: (a) The specific heat C, (b) the in-plane susceptibility χ, and (c) the helicity modulus Υ for the CP^1 model with $\mu = 0$ and $f = 1/2$ as a function of βt. The system size for each curve is $L = 12, 24, 36, 48, 60$. An additional curve $\Upsilon = 8/\pi\beta t$ in (c) indicates the universal jump of a BKT transition.

appropriately adjusting the values of the critical exponents α, ν and T_c. For a finite lattice the peak in the specific heat scales with system size as $C_{\max} \propto L^{\alpha/\nu}$ and occurs at the temperature where the scaling function $\tilde{\phi}(L^{1/\nu}\tilde{t})$ is maximum, defining the finite-lattice transition temperature $T_c(L) = T_c + \text{const} \times L^{-1/\nu}$.

The obtained scaling functions are plotted in Fig. 6 for both the FFXYM and the gauged CP^1 model. For the FFXYM, we obtain $T_c = 0.454 J/k_B$,

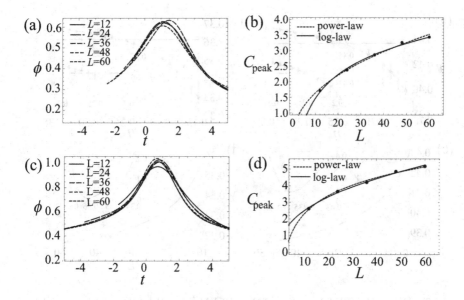

Fig. 6. The left panels show the power-law scaling collapse of the specific heat data for (a) FFXYM and (c) gauged CP^1 model. The right panels show the power-law and logarithmic fitting of the specific heat peak with respect to sizes L for (b) FFXYM and (d) gauged CP^1 model. The power-law fitted value is $\alpha/\nu = 0.439$ for (a) and $\alpha/\nu = 0.397$ for (c).

$\nu = 0.873$ and $\alpha = 0.383$ from Fig. 6(a), which is consistent with the hyperscaling relation $d\nu = 2 - \alpha$ (d is the dimension number). Our scaling analyses support the non-Ising exponent $\nu < 1$, which is consistent with some of the literature.[57–59,61] However, it should be noted that there have been several claims that this non-Ising exponent is caused by the artifact of the finite-size effect and the exponent in the infinite system may be the Ising one $\nu = 1$.[60,64,66] Since our calculation does not have enough system size to resolve this problem and we cannot distinguish the data of Fig. 6(a) as a power-law fitting or logarithmic fitting [see Fig. 6(b)], we shall not go to discuss the details of this issue. Our main claim in this work is that the similar behavior also occurs for the gauged CP^1 model. For this model, we also extract the critical exponents from the same analysis for Fig. 6 (c) and (d) as $T_c = 0.369 J/k_B$, $\nu = 0.878$ and $\alpha = 0.349$. The transition temperature becomes lower than that for the FFXYM, while the critical exponents are similar.

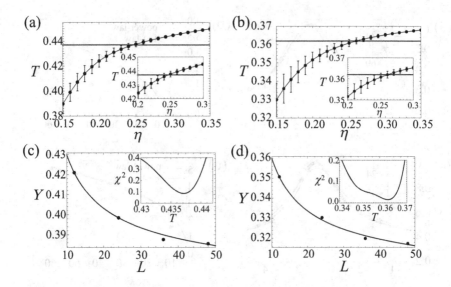

Fig. 7. Data of the BKT transition of the FFXYM [(a) and (c)] and the gauged CP^1 model for $f = 1/2$ [(b) and (d)]. In (a) and (b), we show $T_{\rm BKT}$ obtained by method (i) (see the text) by the horizontal line; (a) $T = 0.437 J/k_B$ and (b) $T = 0.363 J/k_B$. In addition, the temperature corresponding to the crossing point in the method (ii) is plotted as a function of η. Figs. (c) and (d) show the best fit of the scaling relation Eq. (44) at the corresponding $T_{\rm BKT}$, which is obtained by finding the minimum of χ^2-fit error shown in the inset.

Next, we study the U(1)-related phase transition by employing the same analysis for the $f = 0$ case. Some studies revealed that the jump size of Υ at $T = T_{\rm BKT}$ may be non-universal for the FFXYM.[56–59,66] Here, we do not assume $\eta = 1/4$ and evaluate $T_{\rm BKT}$ with two different methods: (i) Using the χ^2-fit of the scaling relation Eq. (44), we evaluate $T_{\rm BKT}$ and the jump size $T^*_{\rm BKT}$ at $T_{\rm BKT}$, which gives the exponent $\eta = T_{\rm BKT}/4T^*_{\rm BKT}$. (ii) We plot $\chi/L^{2-\eta}$ as a function of βt for several system sizes L and take the temperature at the crossing point. We make this analysis by varying η around $1/4$ and search the value of η that gives the same $T_{\rm BKT}$ obtained in the analysis (i). The summary of this analysis is shown in Fig. 7. For the FFXYM, the scaling analysis (i) alone gives $T_{\rm BKT} = 0.437 J/k_B$ and $\eta = 0.2$. This is consistent with the previous literature, where $T_{\rm BKT} = 0.437 J/k_B$ is slightly lower than T_c and the BKT jump is non-universal.[56–59,66] However, the crossing point obtained by the analysis (ii) is preferable to $\eta \approx 0.25$, as shown in Fig. 7(a), which suggests the same universality of the conventional BKT transition. This usual BKT behavior for the FFXYM was also suggested by Olsson.[60] Similar behavior is also found for the gauged CP^1 model as $T_{BKT} = 0.363 J/k_B$ and $\eta \approx 0.22$ for the analysis (i) and $\eta \approx 0.26$ for the analysis (ii).

4.2.3. $f=2/5$

The thermodynamic properties for $f = 2/5$ would appear to be similar to the $f = 1/2$ situation as seen in Fig. 8. However, the nature of the phase transition is very different. In the FXYM, several works indicated that the transition is associated with first-order type.[68,70] Li and Teitel observed hysteresis of the internal energy when the temperature was cycled around the transition and used this as an argument for a first-order transition.[68] Denniston and Tang pointed out that the complicated branching structure of domain walls is similar to the $q > 5$ Pott's models where the first-order transition occurs.[70] The most direct indication of a first-order transition is the presence of a free energy barrier between the ordered and disordered states which diverges as the system size increases.[70] Since there is no diverging characteristic length to which the linear dimension L could be compared at a first-order transition, one finds that it is simply the volume L^2 that controls the size effects.

In the gauged CP^1 model, Fig. 8(a) clearly shows the rapid growth of the peak of C. The inset of Fig. 8(b) shows the peak values of C as a function of L^2. The linear fit clearly shows the expected first-order scaling behavior. From the positions of the peaks as a function of L, we obtain $T_c = 0.193 J/k_B$, which is again slightly lower than that of the FXYM $T_c = 0.2127 J/k_B$.[70] The growing χ and Υ at low temperature certainly provides the emergence of the superfluid order. Due to the presence of the first-order transition, it is difficult to explicitly discuss the properties of the BKT transition.

5. Summary

In this paper, we review the theoretical formulation to treat lattice bosons in an effective magnetic field, especially focusing on the classical regime. The regime described by the XY model (Josephson junction array) can be realized when the mean particle number at each site is very large $\rho_i \gg 1$,[9,35] where the particle number fluctuation becomes negligible as $\sim \rho_i^{-1/2}$. We also study the finite-temperature phase structures of hard-core bosons in a 2D OL subject to an effective magnetic field by employing the gauged CP^1 model. Based on the multicanonical Monte Carlo simulations, we study

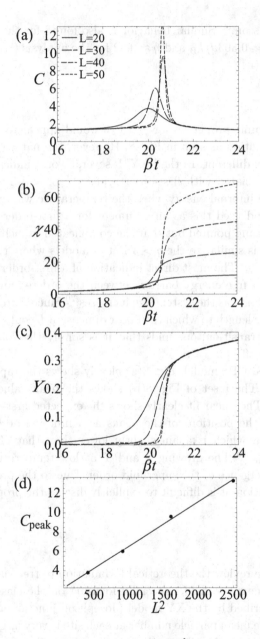

Fig. 8. Several thermodynamic quantities for the CP^1 model with $\mu = 0$ and $f = 2/5$ as a function of βt. (a) The specific heat C, (b) the in-plane susceptibility, and (c) the helicity modulus Υ. The system size for each curve is $L = 20, 30, 40, 50$. (d) represents specific heat vs L^2.

Table 1. List of the transition temperatures and some critical exponents obtained in this work. The transition temperature is measured by using $J = t\rho(1-\rho)$; for $\mu \neq 0$, ρ is used at the corresponding transition temperature. In $\eta(T_{\mathrm{BKT}})$, we represent two values obtained by the method (i) and (ii) in Sec. 4.2.2. For comparison, the corresponding values for the 2D Ising model are $\nu = 1$ and $\alpha = 0$, which implies the logarithmic divergence of the specific heat.

	T_c	ν	α	T_{BKT}	$\eta(T_{\mathrm{BKT}})$
CP^1 model, $f=0$, $\mu=0$	—	—	—	$0.702(1)\ J/k_{\mathrm{B}}$	0.25
CP^1 model, $f=1/2$, $\mu=0$	$0.369(4)\ J/k_{\mathrm{B}}$	0.878(5)	0.349(9)	$0.363(3)\ J/k_{\mathrm{B}}$	(i) 0.22, (ii) 0.26
CP^1 model, $f=2/5$, $\mu=0$	$0.193(1)\ J/k_{\mathrm{B}}$	—	—	—	—
2D XY model	—	—	—	$0.898(1)\ J/k_{\mathrm{B}}$	0.25
2D FFXYM	$0.454(1)\ J/k_{\mathrm{B}}$	0.873(3)	0.383(34)	$0.437(3)\ J/k_{\mathrm{B}}$	(i) 0.2, (ii) 0.25

their phase structures at finite temperatures at half-filling for several values of the magnetic flux per plaquette of the lattice. A summary of the obtained transition temperature and critical exponents are listed in Table 1. The fluctuations of the particle number do not modify the global phase structure and the critical properties of UFXYM. Therefore, the important message of this work is that, even though the strong particle number fluctuation becomes remarkable due to the small site occupation $\rho_i \sim 1$, one can expect similar thermal phase transitions seen in the UFXYM. The recent experimental demonstration on generating an effective magnetic field in an optical lattice[15] opens a door to explore the rich finite-temperature phase diagram of this system. Recently, a simulator of the frustrated classical spin model has been demonstrated using cold atoms in a triangular lattice.[77]

Acknowledgments

I would like to thank Tetsuo Matsui and Yuki Nakano for collaboration and useful discussion. This work is supported in part by a Grant-in-Aid for Scientific Research (Grant No. 21740267) from MEXT, Japan.

Appendix A. Reduction to the Josephson junction regime

Appendix A.1. *Determination of J_{ij}*

Here, we estimate the value of J_{ij} in Eq. (14) based on the simple ansatz of the Wannier function $w_i(\mathbf{r})$ and the experimentally feasible parameters.

We assume that $w_i(\mathbf{r})$ can be decomposed as

$$w_i(\mathbf{r}) = u_i(x,y)v_i(z), \qquad (A.1)$$

where we impose the normalization condition $\int dxdy |u_i(x,y)|^2 = 1$ and $\int dz |v_i(z)|^2 = 1$. The phase fluctuation of the condensates at each well is not important, even for the z-direction, because of the low temperature;[10] both u_i and v_i are thus real functions. We assume that the equilibrium form of Eq. (A.1) is determined by the minimization of the last term of Eq. (14), which is the dominant contribution of the total energy of the system; the other terms describe the dynamics when the system deviates the equilibrium state.

First, we assume that the form of $u_i(x,y)$ is site-independent as $u_i(x,y) = u_0(x-i_x d, y-i_y d)$, where u_0 is located at the origin. Substitution of Eq. (A.1) into the last term of Eq. (14) yields

$$E_{\rm eq} = \sum_i N_i \left\{ \int dxdy u_0 \left[-\frac{\hbar^2}{2m}(\partial_x^2 + \partial_y^2) + V_{\rm OL} \right] u_0 \right.$$
$$\left. + \int dz v_i \left[-\frac{\hbar^2 \partial_z^2}{2m} + V_{\rm ho}(i_x d, i_y d, z) - \mu \right] v_i \right\} + \frac{g}{2} \sum_i N_i^2 \int d\mathbf{r} u_0^4 v_i^4. \quad (A.2)$$

If the laser intensity is large, one can ignore the contribution arising from the two-body interactions for the estimation of $u_0(x,y)$. Then, it is sufficient to minimize the first term of Eq. (A.2). We use the Gaussian ansatz

$$u_0(x,y) = \frac{1}{\sqrt{\pi}\sigma} e^{-(x^2+y^2)/2\sigma^2}, \qquad (A.3)$$

where the variational parameter σ can be obtained by solving numerically $-(q\sigma)^{-4} + se^{-(q\sigma)^2} = 0$ with $s = 2mV_0/\hbar^2 q^2$.

Next, we let $v_i(z)$ have the site dependence. By introducing the site-dependent chemical potential μ_i, the form of $v_i(z)$ is determined by solving

$$\left(-\frac{\hbar^2}{2m}\partial_z^2 + \frac{1}{2}m\omega_z^2 z^2 - \mu_i \right) v_j(z) + g_{\rm 1D} N_i v_i^3 = 0 \qquad (A.4)$$

with $g_{\rm 1D} = g/2\pi\sigma^2$ and

$$\mu_i = \mu - \frac{1}{2}m(\omega^2 - \Omega^2)(i_x^2 + i_y^2)d^2. \qquad (A.5)$$

Here, the ansatz Eq. (A.3) has been used for the last term of Eq. (A.2). We apply the Thomas-Fermi approximation for $v_i(z)$, thus obtaining

$$v_i(z)^2 = \frac{\mu_i}{g_{\rm 1D} N_i}\left(1 - \frac{z^2}{R_{zi}^2}\right) \qquad (A.6)$$

with the Thomas-Fermi radius $R_{zi}^2 = 2\mu_i/m\omega_z^2$ at a site i.

Because of the harmonic confinement, the particle number N_i should be distributed so as to have a peak at the center and to decrease monotonically along the radial direction. We define the index $i_{\max}$ that determines a periphery of the outermost site where the particle number N_i vanishes ($N_i = 0$ for $|i| > i_{\max}$). Then, the chemical potential can be written as $\mu \equiv (m/2)(\omega_\perp^2 - \Omega^2)d^2 i_{\max}^2$, so that

$$\mu_i = \frac{m}{2}(\omega_\perp^2 - \Omega^2)d^2(i_{\max}^2 - i_x^2 - i_y^2). \tag{A.7}$$

Using the normalization condition $\int v_i(z)^2 dz = 1$, we obtain

$$N_i = \frac{4\sqrt{2}}{3}\frac{\mu_i^{3/2}}{g_{1D}\sqrt{m\omega_z^2}}. \tag{A.8}$$

Furthermore, the condition $\sum_i N_i = N$ gives

$$i_{\max} = \frac{a_\perp}{d}\left(\frac{15N}{2\pi}\frac{\omega_z}{\omega_\perp}\frac{ad^2}{a_\perp\sigma^2}\right)^{\frac{1}{5}}\left(1 - \frac{\Omega^2}{\omega_\perp^2}\right)^{-\frac{3}{10}} \tag{A.9}$$

by replacing the summation with the integral $\sum_{i_x,i_y} \to \int di_x di_y$. Here, $a_\perp = \sqrt{\hbar/m\omega_\perp}$ is the characteristic length of the radial harmonic potential. The particle number at each site N_i can have a simple form

$$N_i = \frac{5N}{2\pi i_{\max}^2}\left(1 - \frac{i_x^2 + i_y^2}{i_{\max}^2}\right)^{\frac{3}{2}} \tag{A.10}$$

Using the above results, we can rewrite the parameters in Eq. (14) as

$$J_{ij} \simeq \sqrt{N_i N_j} e^{\frac{-d^2}{4\sigma^2}}\left[\frac{\hbar^2}{2m\sigma^2}\left(\frac{d^2}{4\sigma^2} - 1\right) - V_0 \text{erf}\left(\frac{d}{2\sigma}\right)\right] \tag{A.11}$$

$$E_i = \frac{\hbar^2}{2m\sigma^2} + V_0(1 - e^{-q^2\sigma^2}) - \frac{4}{5}\mu_i \tag{A.12}$$

$$U_i = \frac{3}{10}\frac{g_{1D}}{R_{zi}}, \tag{A.13}$$

where we have used the optimized value of σ and, when calculating the integral in $J_{\mathbf{j},\mathbf{j}'}$, the integral for the z direction was approximated as

$$\int_{-R_{zj}}^{R_{zj}} dz v_i v_j \simeq \sqrt{\int_{-R_{zi}}^{R_{zi}} dz v_i^2 \int_{-R_{zj}}^{R_{zj}} dz v_j^2} \tag{A.14}$$

with Thomas-Fermi radius $R_{zi} \geq R_{zj}$ and the area of the integral for the xy plane as $\int_0^d dx \int_{-\infty}^{+\infty} dy u_0(x,y) V_{\text{OL}} u_0(x,y)$.

Appendix A.2. *Estimation of the parameters*

Following the typical experimental conditions, we estimate the values of the parameters J_{ij}. By assuming ^{87}Rb atoms used in the JILA experiments,[10,13] we use $N = 6 \times 10^5$ and $a = 5.29$ nm. The frequencies of the trapping potential are set as $\omega_\perp = 11.5 \times 2\pi$ and $\omega_z = 50 \times 2\pi$, which gives $a_\perp = 3.2$ μm. In the case of rotation, the range of f is restricted by the harmonic potential because rotation frequency Ω cannot exceed $\omega_\perp$. To prevent the centrifugal expansion at $f = 1/2$, which is the maximum value of f, the condition $d/a_\perp > \sqrt{\pi/2}$ should be satisfied. Thus, the lattice spacing is set as $d = 5$ μm.

For given V_0, we evaluate the variational wave function $u_0(x,y)$ of Eq. (A.3) to obtain the optimized value of σ. Through Eqs. (A.9) and (A.10) with the optimized σ, the value of J_{ij} as well as N_i can be fixed. To see the validity of our model, we first see the density distribution N_i, which is compared with the numerical solution of the 3D Gross-Pitaevskii equation $[-(\hbar^2/2m)\nabla^2 + V_{\text{ex}} + g|\psi|^2]\psi = \mu\psi$. The obtained ground state without rotation ($f = 0$) is shown in Fig. A1(a) and (b). We can see that our model can describe the actual system quantitatively. The particle number at the central well is $N_{(0,0)} \simeq 6000$, decreasing from the center to the outside according to Eq. (A.8). For finite values of f, the centrifugal effect decreases overall N_i, while the region of the occupied sites is expanded.

Next, let us confirm that the conditions of the Josephson regime $J_{ij}/N_i^2 \ll U_i$ and $J_{ij} \gg U_i$ are certainly satisfied. The former condition always holds because of $N_i \gg 1$, even for outermost sites with $N_i \sim 100$. Figure A1(c) represents the V_0 dependence of the ratio J_{ij}/U_i for $f = 0$. The value decreases monotonically with increasing V_0 (for small V_0 the tight binding approximation may be broken), quantitatively similar to the estimation by Trombettoni *et al.* using a simple Gaussian ansatz of $w_i(\mathbf{r})$ for all directions.[9] For the central region $(i_x, i_y) = (0,0)$ $(j_x, j_y) = (1,0)$ the condition $J_{ij} \gg U_j$ is well satisfied for $30 \leq V_0/\hbar\omega_\perp \leq 90$. Even near the edge of the condensate in the xy plane, where the particle number is less than that in the center, the condition $J_{ij} \gg U_i$ is still good; for example, the case for $(i_x, i_y) = (5,0)$ $(j_x, j_y) = (6,0)$ is shown in Fig. A1(a). Therefore, the quantum correction arising from the second term of Eq. (14) may be neglected in the typical experiment of the trapped BECs.

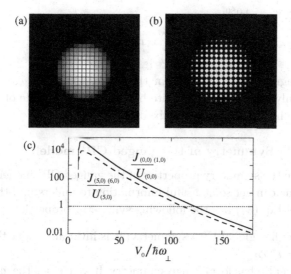

Fig. A1. (a) The distribution of the particle number N_i in a 2D OL, obtained by Eq. (A.10). (b) The result of the numerical solution of the 3D Gross-Pitaevskii equation with relevant values of the parameters. The figure shows the density profile integrated along the z-axis $n(x,y) = \int dz |\psi(x,y,z)|^2$. The parameter values used are $N = 6 \times 10^5$, $a = 5.29$ nm, $\omega_\perp = 11.5 \times 2\pi$, $\omega_z = 50 \times 2\pi$, $d = 5$ μm and $V_0 = 65\hbar\omega_\perp$. Both plots have the same relative density scale. (c) J_{ij}/U_i as a function of the optical lattice depth V_0.

Appendix B. Relation between the CP^1 model and the other models

If we take into account the nearest-neighbor repulsive interaction in the Bose-Hubbard model, the hard-core constraint yields the gauged spin-half quantum XXZ model[78]

$$\hat{H}_{\text{XXZ}} = -\frac{t}{2} \sum_{\langle i,j \rangle} \left(\hat{s}_i^+ \hat{s}_j^- e^{iA_{ij}} + \text{c.c.} \right) + V \sum_{\langle i,j \rangle} \hat{s}_i^z \hat{s}_j^z - \mu \sum_i \hat{s}_i^z. \quad \text{(B.1)}$$

Our model corresponds to the classical version of the gauged XXZ model with $V = 0$, called the XX0 (three-component XY) model.[79] The XX0 model is clearly distinct from the XY (two-component XY) model, as the spins fluctuate also out of the xy plane. In other words, even if the XX0 model and the XY model share the same form of the Hamiltonian, the associated phase space is different.

In addition, we note that there is a similar model in which the spins are random in not only their direction but also their magnitudes, known as the "fuzzy" spin XY model[80]

$$H_{\text{FXY}} = -J \sum_{\langle i,j \rangle} x_i x_j \cos(\theta_i - \theta_j). \tag{B.2}$$

While the magnitudes x_i of spins in the fuzzy XY model can have any values randomly and continuously site by site, the magnitude of O(3) spin $\vec{s}_i$ at each site in the CP^1 model is fixed unity.

Appendix C. Symmetry of the gauged CP^1 model

We summarize the symmetry properties of the gauged CP^1 model Eq. (34), from which one can get some useful information to understand the results.

The model Eq. (34) has the following symmetry properties:

(1) The model has global U(1) symmetry; it is invariant under the change $\phi_i \to \phi'_i = e^{i\theta}\phi_i$.
(2) The model also has local gauge symmetry. If we change the gauge $\mathbf{A} \to \mathbf{A} + \nabla \chi$, then the Hamiltonian remains unchanged if the boson picks up a phase change as $A_{ij} \to A_{ij} + (\chi_j - \chi_i)$ and $\phi_i \to e^{i\chi_i}\phi_i$. In terms of the pseudospin $\vec{s}_i$ for the i-th site, this corresponds to a rotation with the angle χ_i in the xy plane, $s_i^\pm \to e^{\mp i\chi_i}s_i^\pm$. Because the choice $\chi_i = \pi i_x i_y$ gives rise to a shift of fluxes $f \to f+1$, the system is periodic in f with the period 1.
(3) The Hamiltonian Eq. (34) is invariant under the change $f \to f' = -f$, corresponding to the time reversal operation $\phi_i \to \phi' = \phi_i^*$ ($\lambda_{ai} \to \lambda'_{ai} = -\lambda_{ai}$). Since the partition function is not affected by this transformation, one can show that the internal energy $E = \langle H \rangle = -\partial \ln Z/\partial \beta$ and the mean number density $\rho = \langle N \rangle/L^2 = L^{-2}\partial \ln Z/\partial(\beta\mu)$ have the following properties

$$E(\beta, t, \mu, f) = E(\beta, t, \mu, -f), \tag{C.1}$$

$$\rho(\beta, t, \mu, f) = \rho(\beta, t, \mu, -f). \tag{C.2}$$

The symmetric form of the ground state energy shown in Fig. 1 can be understood as follows. The plotted energy $E_{\min}$ in Fig. 1 is the first term of Eq. (34) $E = E_{\min} - \mu\rho L^2$. From Eq. (C.2), one can see $\mu(\rho(f), \beta, t, f) = \mu(\rho(-f), \beta, t, -f)$. For the fixed ρ, Eq. (C.1) gives

$$\begin{aligned} E_{\min}(t, \mu(f), f) &= E_{\min}(t, \mu(-f), -f) \\ &= E_{\min}(t, \mu(1-f), 1-f). \end{aligned} \tag{C.3}$$

The last equality is due to the invariance for $f \to f + 1$.

(4) Let us consider the transformation $\psi_i \to \psi'_i = \pi - \psi_i$. Then, Eq. (36) yields $H_{\mathrm{CP}^1}(\mu) \to H'_{\mathrm{CP}^1}(\mu) = H_{\mathrm{CP}^1}(-\mu)$ and $\rho = \langle \cos^2(\psi/2) \rangle \to \rho'(\mu) = 1 - \rho(-\mu)$ (or equivalently $s_i^z \to s_i^{\prime z} = -s_i^z$). This reflects the particle-hole symmetry, where ρ' should be interpreted as the hole density if ρ represents the particle density. Therefore, especially for $\mu = 0$, we obtain the relation

$$\rho(\beta, t, 0, f) = \rho'(\beta, t, 0, f) = 1 - \rho(\beta, t, 0, f) = \frac{1}{2}. \qquad (\mathrm{C}.4)$$

References

1. M. Lewenstein, A. Sanpera, V. Ahufinger, B. Damski, A. Sen De, U. Sen, *Adv. Phys.* **56**, 243 (2007).
2. I. Bloch, J. Dalibard, and W. Zwerger, *Rev. Mod. Phys.* **80**, 885 (2008).
3. D. Jaksch, C. Bruder, J.I. Cirac, C.W. Gardiner, and P.Zoller, *Phys. Rev. Lett.* **81**, 3108 (1998).
4. M. Greiner, O. Mandel, T. Esslinger, T.W. Hänsch, and I. Bloch, *Nature (London)* **415**, 39 (2002).
5. M. P. A. Fisher, P. B. Weichman, G. Grinstein, and D. S. Fisher, *Phys. Rev. B* **40**, 546 (1989).
6. B.P. Anderson and M.A. Kasevich, *Science*, **282**, 1686 (1998).
7. F.S. Cataliotti, S. Burger, C. Fort, P. Maddaloni, F. Minardi, A. Trombettoni, A. Smerzi, and M. Inguscio, *Science*, **293**, 843 (2001).
8. R. Fazio and H. van der Zant, *Phys. Rep.* **355**, 235 (2001).
9. A. Trombettoni, A. Smerzi, and P. Sodano, *New J. Phys.* **7**, 57 (2005).
10. V. Schweikhard, S. Tung, and E. A. Cornell, *Phys. Rev. Lett.* **99**, 030401 (2007).
11. J. Dalibard, F. Gerbier, G. Juzeliūnas, and P. Öhberg, *Rev. Mod. Phys.* **83**, 1523 (2011).
12. N. R. Cooper, *Adv. Phys.* **57**, 539 (2008).
13. S. Tung, V. Schweikhard, and E.A. Cornell, *Phys. Rev. Lett.* **97**, 240402 (2006).
14. R. A. Williams, S. Al-Assam, C. J. Foot, *Phys. Rev. Lett.* **104**, 050404 (2010).
15. M. Aidelsburger, M. Atala, S. Nascimbène, S. Trotzky, Y.-A. Chen, I. Bloch, *Phys. Rev. Lett.* **107**, 255301 (2011).
16. D. Jaksch and P. Zoller, *New J. Phys.* **5**, 56 (2003).
17. E. J. Mueller, *Phys. Rev. A* **70**, 041603(R) (2004).
18. A. S. Sørensen, E. Demler, and M. D. Lukin, *Phys. Rev. Lett.* **94**, 086803 (2005).
19. F. Gerbier and J. Dalibard, *New J. Phys.* **12**, 033007 (2010).
20. M. Ö. Oktel, M. Nita, and B. Tanatar *Phys. Rev. B* **75**, 045133 (2007).
21. R. O. Umucalilar and M. Ö. Oktel *Phys. Rev. A* **76**, 055601 (2007).
22. D. S. Goldbaum and E. J. Mueller, *Phys. Rev. A* **77**, 033629 (2008); *ibid* **79**, 021602 (2009).
23. R. N. Palmer, A. Klein, and D. Jaksch, *Phys. Rev. A* **78**, 013609 (2008).

24. T. P. Polak and T. K. Kopeć Phys. Rev. A **79**, 063629 (2009).
25. G. Möller and N. R. Cooper, Phys. Rev. Lett. **103**, 105303 (2009).
26. R. O. Umucalilar and E. J. Mueller Phys. Rev. A **81**, 053628 (2010) .
27. T. Durić and D. K. K. Lee, Phys. Rev. B **81**, 014520 (2010).
28. H. Zhai, R. O. Umucalilar, and M. Ö. Oktel Phys. Rev. Lett. **104**, 145301 (2010).
29. S. Powell, R. Barnett, R. Sensarma, and S. Das Sarma Phys. Rev. Lett. **104**, 255303 (2010), Phys. Rev. A **83**, 013612 (2011).
30. G. Möller and N. R. Cooper, Phys. Rev. A **82**, 063625 (2010).
31. D. R. Hofstadter, Phys. Rev. B **14**, 2239 (1976).
32. E. Brown, Phys. Rev. **133**, A1038 (1964).
33. J. Zak, Phys. Rev. **134**, A1602 (1964); **134**, A1607 (1964).
34. M. Polini, R. Fazio, A. H. MacDonald, and M. P. Tosi, Phys. Rev. Lett. **95**, 010401 (2005).
35. K. Kasamatsu, Phys. Rev. A. **79**, 021604(R) (2009).
36. S. Teitel and C. Jayaprakash, Phys. Rev. Lett. **51**, 1999 (1983).
37. T. C. Halsey, Phys. Rev. B **31**, 5728 (1985).
38. J. P. Straley and G. M. Barnett, Phys. Rev. B **48**, 3309 (1993).
39. T.C. Halsey, Phys. Rev. Lett. **55**, 1018 (1985).
40. B. Kim and S.J. Lee, Phys. Rev. Lett. **78**, 3709 (1997).
41. C. Denniston and C. Tang, Phys. Rev. B **60**, 3163 (1999).
42. C. Chin, R. Grimm, P. Julienne, and E. Tiesinga, Rev. Mod. Phys. **82**, 1225 (2010).
43. J. Ruseckas, G. Juzeliunas, P. Ohberg, and M. Fleischhauer, Phys. Rev. Lett. **95**, 010404 (2005).
44. S. L. Zhu, H. Fu, C. J. Wu, S. C. Zhang, and L. M. Duan, Phys. Rev. Lett. **97**, 240401 (2006).
45. X. J. Liu, X. Liu, L. C. Kwek, and C. H. Oh, Phys. Rev. Lett. **98**, 026602 (2007).
46. K. J. Gunter, M. Cheneau, T. Yefsah, S. P. Rath, and J. Dalibard, Phys. Rev. A **79**, 011604 (2009).
47. I. B. Spielman, Phys. Rev. A **79**, 063613 (2009).
48. Y.-J. Lin, R. L. Compton, A. R. Perry, W. D. Phillips, J. V. Porto, and I. B. Spielman Phys. Rev. Lett. **102**, 130401 (2009).
49. Y. Lin, R. L. Compton, K. J. Garcia, J. V. Porto, and I. B. Spielman, Nature (London) **462**, 628 (2009).
50. Y-J. Lin, R. L. Compton, K. Jiménez-García, W. D. Phillips, J. V. Porto and I. B. Spielman, Nat. Phys. **7**, 531 (2011).
51. Y.-J. Lin, K. Jiménez-García and I. B. Spielman, Nature (London) **471**, 83 (2011).
52. Z. Fu, P. Wang, S. Chai, L. Huang, and J. Zhang, Phys. Rev. A **84**, 043609 (2011).
53. Y. Nakano, T. Ishima, N. Kobayashi, K. Sakakibara, I. Ichinose, T. Matsui Phys. Rev. B **83**, 235116 (2011).
54. A. Shimizu, K. Aoki, K. Sakakibara, I. Ichinose, and T. Matsui, Phys. Rev. B **83**, 064502 (2011).

55. M. Campostrini and E. Vicari, *Phys. Rev. Lett.* **102**, 240601 (2009).
56. S. Teitel and C. Jayaprakash, *Phys. Rev. B* **27**, 598 (1983).
57. J. Lee, J. M. Kosterlitz and E. Granato, *Phys. Rev. B* **43**, 11531 (1991).
58. G. Ramirez-Santiago and J. V. José, *Phys. Rev. B* **49**, 9567 (1994).
59. S. Lee and K.-C. Lee, *Phys. Rev. B* **49**, 15184 (1994).
60. P. Olsson, *Phys. Rev. Lett.* **75**, 2758 (1995).
61. H.J. Luo, L. Schülke, and B. Zheng, *Phys. Rev. Lett.* **81**, 180 (1998).
62. S. E. Korshunov, *Phys. Rev. Lett.* **88**, 167007 (2002).
63. M. Hasenbusch, A. Pelissetto, and E. Vicari, *Phys. Rev. B* **72**, 184502 (2005).
64. P. Olsson and S. Teitel, *Phys. Rev. B* **71**, 104423 (2005).
65. P. Minnhagen, B.J. Kim, S. Bernhardsson, and G. Cristofano, *Phys. Rev. B* **78**, 184432 (2008).
66. S. Okumura, H. Yoshino, and H. Kawamura, *Phys. Rev. B* **83**, 094429 (2011).
67. G. S. Grest, *Phys. Rev. B* **39**, 9267 (1989).
68. Y.-H. Li and S. Teitel, *Phys. Rev. Lett.* **65**, 2595 (1990).
69. S. Lee and K.-C. Lee, *Phys. Rev. B* **52**, 6706 (1995).
70. C. Denniston and C. Tang, *Phys. Rev. B* **58**, 6591 (1998).
71. B. A. Berg, *Fields Inst. Commun.* **26**, 1 (2000).
72. D. R. Nelson and J. M. Kosterlitz, *Phys. Rev. Lett.* **39**, 1201 (1977).
73. H. Weber and P. Minnhagen, *Phys. Rev. B* **37**, 5986 (1988).
74. P. Minnhagen, *Phys. Rev. Lett.* **54**, 2351 (1985).
75. R. Gupta, J. DeLapp, G. G. Batrouni, G. C. Fox, C. F. Baillie, and J. Apostolakis, *Phys. Rev. Lett.* **61**, 1996 (1988).
76. See, e.g., *Finite Size Scaling and Numerical Simulation of Statistical Systems*, ed. V. Privman (World Scientific, Singapore, 1990) and references cited therein.
77. J. Struck, C. Ölschläger, R. Le Targat, P. Soltan-Panahi, A. Eckardt, M. Lewenstein, P. Windpassinger, K. Sengstock, *Science* **333** 996 (2011).
78. N. H. Lindner, A. Auerbach, and D. P. Arovas, *Phys. Rev. Lett.* **102**, 070403 (2009), *Phys. Rev. B* **82**, 134510 (2010).
79. A. Cuccoli, V. Tognetti, and R. Vaia, *Phys. Rev. B* **52**, 10221 (1995).
80. T. Kawasaki and S. Miyashita, *Prog. Theor. Phys.* **93**, 47 (1995).

QUANTUM SIMULATION USING EXCITON-POLARITONS AND THEIR APPLICATIONS TOWARD ACCELERATED OPTIMIZATION PROBLEM SEARCH

TIM BYRNES
KAI YAN
KENICHIRO KUSUDO
MICHAEL FRASER

National Institute of Informatics, Chiyoda-ku, Tokyo, 101-8430, Japan

YOSHIHISA YAMAMOTO

National Institute of Informatics, Chiyoda-ku, Tokyo, 101-8430, Japan
E. L. Ginzton Laboratory, Stanford University, Stanford, CA 94305, USA

Exciton-polaritons in semiconductor microcavities have been recently observed to undergo dynamical Bose-Einstein condensation. This opens the door for many of the analogous experiments to be performed on a semiconductor chip that have been performed in atomic systems. One of the future applications of quantum technology is quantum simulation, where quantum many body problems are directly simulated in the laboratory. Recent advances towards the realization of quantum simulators using exciton-polaritons will be reviewed. In particular, we discuss in detail one application of the ideas of quantum simulation, where Bose-Einstein condensates are used to accelerate the speed of finding the ground state of a given optimization problem. The method uses the principle of indistinguishability, in contrast to the principle of superposition which is traditionally used to obtain a quantum speedup. By taking advantage of bosonic final state stimulation, the convergence to the ground state can be accelerated by a factor of N, where N is the number of bosons in the BEC.

1. Introduction

Quantum simulation has attracted a large amount of attention in recent years[1]. The aim of quantum simulation is to give another way to investigating difficult quantum many-body problems by simulating quantum models experimentally on other quantum devices. This is advantageous because it is often easier to control the model parameters in the artificially fabricated device, rather than the original system being examined. Currently, most researchers working on quantum simulation have been working with cold atom systems[2-5] and ion trap systems[6,7]. We have taken an alternative system for performing quantum simulations: exciton-polaritons condensates. In this paper, we briefly introduce works relating to quantum simulation with exciton-polaritons. In section 2, we give a review of the fundamentals of quantum simulation using one of the most studied models in quantum simulation: the Hubbard model. In section 3, we review recent developments achieving exciton-polaritons condensation. In section 4, we describe some key topics relating to quantum simulation with exciton-polaritons. An application of the quantum simulation approach to solving optimization problems using BECs is desicussed in section 5. Finally, we summarize this paper and give future prospects in section 6.

2. Quantum Simulation of the Hubbard Model

The Hubbard model is particularly interesting as it is one of the standard models used to describe strongly correlated phenomena such as metal-insulator transition[8], magnetism[9], and high temperature superconductivity[10]. Hubbard models are constructed on discrete lattice systems. Depending on whether the particles in the Hubbard model are bosons or fermions, we have either a Bose Hubbard model or a Fermi Hubbard model. The form of the Hamiltonian of these two types of Hubbard model is

$$H_{Bose} = -t \sum_{<i,j>} \hat{a}_i^\dagger \hat{a}_j + \sum_i \varepsilon_i \hat{n}_i + \frac{1}{2} U \sum_i \hat{n}_i (\hat{n}_i - 1) \tag{1}$$

and

$$H_{Fermi} = -t \sum_{<i,j>,\sigma} \hat{c}_{j\sigma}^\dagger \hat{c}_{i\sigma} + \sum_i \varepsilon_i \hat{n}_i + U \sum_i \hat{n}_{i\uparrow} \hat{n}_{i\downarrow} \tag{2}$$

respectively. Here, t is the hopping matrix element between nearest neighbor sites, U is the interaction energy for two particles existing on the same site, σ labels the spin, and a ($a^\dagger$) and c ($c^\dagger$) is the annihilation (creation) operator of bosons and fermions respectively. Summation is taken over all sites i, and all combination of nearest neighbor sites $<i, j>$. The difference between these two models comes from the difference between the statistics of bosons and fermions. For bosons, any number of particles can occupy the same state (or site in this case). For fermions, due to the Pauli exclusion principle, only two fermions (of up-spin and down-spin) can enter the same state.

In both Hamiltonians, the first term represents the kinetic energy, the second represents the energy off-set on each site, and the last represents the repulsive

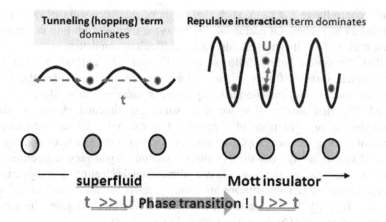

Figure 1. Schematic picture of the phase transition in the Hubbard models.

interaction between two particles. Hubbard models capture the competition between the kinetic energy and the on-site interaction energy. At some critical value in the ratio of U/t, a quantum phase transition from non-interacting delocalized states to Mott-insulator is known to occur (see Figure 1). To see that such a transition must occur, let us assume zero temperature and a constant energy offset $\varepsilon_i = 0$ in the following discussion. First, when the kinetic term is dominant (corresponding to the $t \gg U$ limit), the ground state can be approximated by the first term only, corresponding to either free fermions or bosons, depending on the particle type. In bosonic systems with a large number of particles, this limit corresponds to a Bose-Einstein condensed state, since all the bosons all occupy the same zero momentum quantum state. For fermionic systems, we have a Fermi sea, corresponding to a conducting metallic state. Now consider the opposite limit, when the on-site interaction is dominant ($t \ll U$). Due to the repulsive interaction, particles cannot tunnel into to neighboring sites if it is occupied. The minimized interaction energy is then realized when the particle number per a site is fixed to the average number N/M, where N is the number of particles and M is the number of sites. This is called Mott-insulator phase, because the tunneling is suppressed by the repulsive interaction and the system behaves as an insulator.

We now turn to how a quantum simulation of such a Hubbard model may be realized in the laboratory. The aim of quantum simulation is to create an artificial quantum system such that the desired Hamiltonian is created and controlled. In the case of the Hubbard model, we would like to realize a system with a controllable U/t ratio, such that some of the interesting physics in the system can be probed. Of particular interest is the region exhibiting the quantum phase transition, as it is numerically rather difficult to simulate.

The discussion below follows Refs. 12 and 13, and we assume one-dimensional systems for simplicity. Moreover we only consider the Fermi Hubbard model here, but the basic argument is same as that of Bose Hubbard model. We first assume that we have some experimental system available containing a large number of fermions. For example, for atomic systems, fermionic atoms can be trapped in magneto-optical traps and laser cooling can be used to efficiently cool the system to quantum degeneracy temperatures[4,5]. Another example is the use of two dimensional electron gas systems in semiconductors[12,13]. The fermions are assumed to follow a known interaction of the form $U(x,x')$. For example, for two dimensional electron gases the form of the interaction is the Coulomb interaction $U(x,x') = e^2/4\pi\varepsilon|x-x'|$.

To model such a system, we now apply a periodic potential, which can be taken to be of the form $V(x) = W_0\sin(kx)$, where k is the wave number of the periodic potential and x is the real coordinate. This can be done for atomic systems by creating an optical lattice, which consists of standing wave laser fields, trapping atoms at the antinodes. For the two dimensional electron gas system, periodic metal patterning with an applied voltage on the surface can produce similar periodic potentials. The total Hamiltonian of such a system in second quantized form is

$$H = \sum_\sigma \int dx \psi_\sigma^\dagger(x) \left[-\frac{\hbar^2}{2m}\nabla^2 + V(x) \right] \psi_\sigma(x)$$
$$+ \frac{1}{2} \sum_{\sigma\sigma'} \int dx dx' \psi_{\sigma'}^\dagger(x') \psi_\sigma^\dagger(x) U(x,x') \psi_\sigma(x) \psi_{\sigma'}(x') \qquad (3)$$

where σ and σ' denotes the spin of fermions, ψ and $\psi^\dagger$ are the fermionic field operators and its conjugate. $\hbar$ is the reduced Plank constant, m is the particle mass. Due to the periodic potentials, a band structure is formed in the system. If the particle number is enough small and the potential is deep enough that only the first band is occupied, the Wannier basis can be used to decompose the field operator:

$$\psi_\sigma(x) = \sum_i w_i(x-x_i) \hat{c}_{i\sigma} . \qquad (4)$$

Here $w_i(x-x_i)$ is the Wannier function of the first band on site i. Substituting (4) into (3), we obtain

$$H = \sum_{ij,\sigma} t_{ij}\left(\hat{c}_{j\sigma}^\dagger \hat{c}_{i\sigma} + \hat{c}_{i\sigma}^\dagger \hat{c}_{j\sigma}\right) + \sum_{i,\sigma} \varepsilon_i \hat{n}_{i\sigma} + \sum_{ijkl,\sigma\sigma'} U_{ijkl} \hat{c}_{i\sigma}^\dagger \hat{c}_{j\sigma'}^\dagger \hat{c}_{k\sigma'} \hat{c}_{l\sigma'} \qquad (5)$$

where $n_{i\sigma} = c^{\dagger}_{i\sigma} c_{i\sigma}$. t_{ij} and U_{ijkl} are written in Ref. 12 (Eq. 4 and 5). If we consider $t_{ij} \neq 0$ only when i and j are nearest neighbor sites, and $U_{ijkl} \neq 0$ only when $i = j = k = l$, then the Hamiltonian of Eq. (5) reduces to that of Eq. (2), and we obtain the Fermi Hubbard Hamiltonian.

By using the above technique to simulate the Hubbard model, a quantum phase transition between a Bose-Einstein condensate (BEC) phase and a Mott insulator phase in the Bose Hubbard model was observed experimentally in Ref. 3. This was followed by a similar quantum phase transition in the Fermi Hubbard model in Ref. 4, 5. The way that the transition was observed was by changing the depth of the potential W_0 of the optical lattice creating the periodic potential. This allows for the key parameter U/t to be swept widely. In both cases, the atoms are prepared in a cooled state and then the optical lattice is induced adiabatically. In the bosonic system, the loss of the phase coherence over the entire system due to the formation of Mott insulator states indicated the signature of the phase transition[3]. In the fermionic system, the suppression of doubly occupied sites and gapped excitation modes were detected, both of which are typical signatures of the Mott insulator state[4].

3. Exciton-Polaritons

In this section, we explain the system that we plan to use for quantum simulation, and briefly introduce our previous work with exciton-polariton condensation. Microcavity exciton-polaritons (EPs) are quasi-particles resulting from the strong coupling between quantum well (QW) excitons and cavity photons. An exciton is an excitation mode of the semiconductor crystal ground state, a bound state of a free electron and a free hole. Because excitons have a dipole moment, they can couple with photon modes whose energy is close to the exciton energy. When the strength of the coupling is larger than the line-width of the exciton, which is due to the spontaneous emission and collisional broadening, two new eigenstates appear. These are known as the upper (exciton) polariton (UP) and lower (exciton) polariton (LP) modes for the high and low energy branches respectively. Although the growth axis translational symmetry is broken due to presence of the 2D-confinement in the QW layer, the in-plane translational symmetry is conserved. Therefore photons and excitons with the same in-plane momentum $k_{\parallel}$ are coupled, and LP and UP modes with a characteristic dispersion in $k_{\parallel}$ appear. After the first observation of the strong coupling between QW excitons and cavity photons[14], EPs have been intensively investigated. One of the most intriguing topics is the condensation of EPs[15]. The EPs system (particularly the LP) has long been considered a promising candidate for BEC in solids[16] because of the extremely low LP effective mass and associated high condensate transition temperature.

The EP effective mass is around four orders of magnitude lighter than that of an exciton, and eight orders of magnitude lighter than that of atoms. This is advantageous for BEC, because the critical temperature of BEC is inversely proportional to the particle mass. The mass of EP m_{LP} is found through the weighted mixture of the exciton and photon masses, by the following equation,

$$\frac{1}{m_{LP}} = \frac{|X|^2}{m_{ex}} + \frac{|C|^2}{m_{ph}} \qquad (6)$$

where m_{ex} and m_{ph} are the exciton and cavity photon masses respectively. X and C are the Hopfield coefficients which are normalized coefficients describing the fraction of exciton and cavity photon in the EP coherent mixture. This is defined through the annihilation operator a_{LP} of EP, which is written in terms of the exciton a_{ex} and cavity photon a_{ph} operators as $a_{LP} = Xa_{ex} + Ca_{ph}$. $|X|^2$ and $|C|^2$ for LPs are then given by Eqs. 7 and 8 respectively,

$$|X|^2 = \frac{1}{2}\left(1 + \frac{\Delta}{\sqrt{\Delta^2 + 4\hbar^2\Omega^2}}\right) \quad |C|^2 = \frac{1}{2}\left(1 - \frac{\Delta}{\sqrt{\Delta^2 + 4\hbar^2\Omega^2}}\right) \qquad (7)$$

where $\Delta = E_{cav}(k_\parallel = 0) - E_{ex}(k_\parallel = 0)$ is the detuning parameter, and $2\hbar\Omega$ is the Rabi splitting energy describing the coupling strength of exciton and cavity photon, or equivalently the minimum energy separation between LP and UP modes, where $E_{cav}(k_\parallel = 0)$ and $E_{ex}(k_\parallel = 0)$ are the cavity photon energy and exciton energy at $k_\parallel = 0$. Equation (7) is only valid when the Rabi splitting is significantly larger than the detuning. The condition where $\Delta > 0$ is called the 'Red detuning' regime, and $\Delta < 0$ the 'Blue detuning' regime, as in the experiments it is typically the cavity photon resonance energy which is varied to adjust the detuning. Polaritons are easiest to produce when the cavity photon mode is resonant with the exciton mode, known as the zero detuning condition ($\Delta = 0$), at which point the LP mode is an even half exciton-half photon coherent mixture. When the detuning is reduced into the red detuning regime, the LP mode (lowest energy eigenmode) becomes more photon like and lower in mass, whereas increasing the detuning into the blue detuned regime gives a more exciton-like LP and a consequently heavier mass. The exciton energy is fixed by the choice of the QW material and structure and it is simplest to control the detuning parameter by changing the cavity resonance. A semiconductor microcavity is created with top and bottom distributed Bragg reflectors (DBR), the resonance energy of which is determined by the thickness of the alternating low and high dielectric permittivity layers. Essentially, we can relatively easily control the LP effective mass through variation of the detuning parameter. Figure 2 shows the exciton fraction dependence of the LP mass. To achieve any significant variation in the effective mass however, as clear in Fig. 2b one must consider the detuning region of greater than about 95% exciton fraction.

Recently several groups[17-20] have claimed observation of LPs condensation. Other phenomena related to superfluidity have also been reported, such as formation of a quantized vortex[21], and frictionless flow[22].

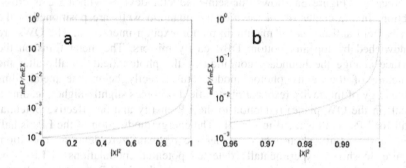

Figure 2. (a) Dependence of the LP effective mass m_{LP} on the fraction of exciton composing the polariton. (b) The dependency is heavily biased toward the large exciton fraction ($|X|^2 > 0.96$). Here we assume $m_{ph} = 10^{-4} \times m_{ex}$.

4. Quantum Simulation with Exciton-Polaritons

Our motivations for quantum simulation with EPs are as follows.

1. Investigating models such as Hubbard models in several systems is important, because those models are presently unsolvable both analytically and numerically. Thus only through a comparison of different quantum simulators can one gain confidence in the results.
2. The EPs are directly and easily accessible to measurement through the leaked photons from semiconductor cavities. The EP state can thus be easily studied with methods well established in quantum optics.
3. Due to the peculiar and highly tunable properties of the EP (very large parameter space), we expect to observe new physics in EP artificial-atom systems.

Here we introduce our theoretical and experimental work investigating quantum simulation with EPs.

4.1. *Excited state condensation in one dimensional periodic lattice potentials*

As is shown in section II, periodic lattice structures are crucial for the Hubbard models. Periodically modulated lattice potentials have been implemented in EP systems by depositing and patterning thin metal films on the surface of the microcavity. Figure 3a shows our semiconductor devices with a z-axis cross-section. In the center of the cavities the quantum wells are positioned at the cavity field anti-nodes for maximum photon-exciton interaction. The QWs are sandwiched by top and bottom DBR cavity mirrors. The metal films on the surface change the boundary condition of the photon field (locally alter the structure of the cavity photon mode), thus directly below the metal films the energy of the cavity resonant photon field becomes slightly higher, leading a local (in the QW plane) reduction to the EP energy and an effective potential wall for EPs, as illustrated in Fig. 3b. The energy modulation of the LPs is half that of the cavity photon mode under the resonant zero-detuning condition. Figure 3c shows experimentally detected potential modulations to LPs. The presence of the metal films spatially modulates the the LP energy by about a hundred μeV.

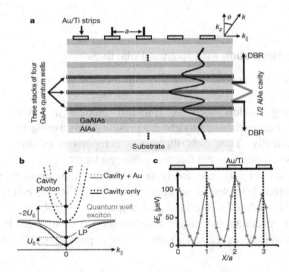

Figure 3. (from Ref. 23) (a) Schematic of sample cross-section for EP potential modulation with patterned metallic surface layers, (b) Effective resulting shift to the cavity and LP energies and an experimentally measured spatial modulation of the LP energy.

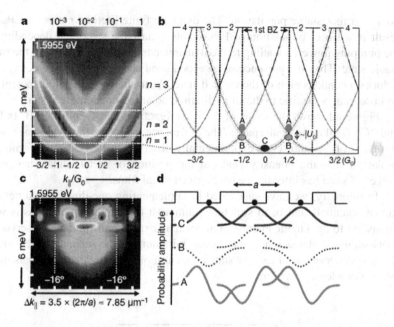

Figure 4. (from Ref. 23) Band structures and excited state condensation of EPs.

Figure 4a shows the observed LP dispersion below the threshold pump power with the 1D periodic potentials described in Fig. 3, and Fig. 4b is the corresponding band structure. In Fig. 4a, b, c, the horizontal axis is the LP wave number normalized by $G_0 = 2\pi/\lambda$ where λ is the period of the potentials and the vertical axis is the LP energy. U_0 is the trapping potential depth, and n denotes the band index. In Fig. 4a we can see two weaker dispersion replicas of the extended Brillouin zones (BZ) in addition to the bright central one at $k_\parallel = 0$ (1st BZ) due to the periodic lattice potentials. The ground state of the system is located on point 'C' in Fig. 4b, but we also observed condensation at the 'A' point which is the energy minimum in the second lowest band. Figure 4c shows the observed EP dispersion above the threshold pump power, clearly showing condensation at both the 'C' point (the lower energy and at $k_\parallel = 0$) and 'A' point (the higher energy two at $k_\parallel/G_0 = \pm 1/2$).

The explanation is as follows. In semiconductor devices, EPs are optically injected with large momentum which are subsequently cooled down due to scattering with phonons and EPs. In the presence of a band structure such as the one described, the decay from a higher band to a lower band is inhibited with some characteristic decay rate, so that the higher band energy minimum can possess a metastable population. The dynamic finite-lifetime nature of EPs can result in a larger population of EPs in this metastable state, and depending on the pumping power, a leading to the dynamic condensation of $k_{\parallel} \neq 0$ at the metastable state. Thus the presence of a gapped band structure and finite decay rates induces a condensation of the excited state 'A' in the periodic lattice potentials, an observation specific to the finite-lifetime polariton condensate system.

Figure 4d shows the shape of the wave functions at 'A' (p-like), 'B' (s-like) and 'C' (s-like) in real space. The difference between the real space wave functions of 'A' and 'C' is experimentally observed in Ref. 23, specifically, the p-like 'A' has the intensity peaks at the position of metal films whereas the s-like 'C' state has intensity peaks between metal films.

In summary, we have succeeded in implementing one-dimensional periodic lattice potentials and band structures, although the depth of potentials is weak compared to the kinetic energy. With such an experiment, we have observed the meta-stable condensation at the energy minimum of the second band, which is the first observation of excited state condensation in the lattice potentials in any present condensate system.

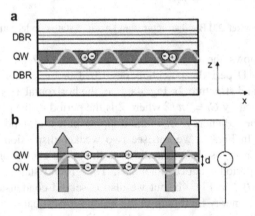

Figure 5. (from Ref. 24) Schematic picture of an usual polariton system and indirect exciton system.

4.2. *Mott transition of EPs and indirect excitons in a periodic potential*

In order to realize a Mott insulator state, we must know what experimental parameters are optimal for achieving this goal. The work in Ref. 24 gives a detailed study of the Hubbard parameters t and U for EP systems as a function of the lattice potential strength. One of the difficulties for quantum simulation of

Hubbard models with EPs is that the EP mass is so light that the kinetic energy is usually much bigger than the EP-EP interactions. The interactions originate principally from the Coulomb interaction of EPs excitonic component. To increase the Hubbard parameter U, we estimated the effect of stronger trapping and heavier EPs. The former would be achieved by deepening the potentials, the latter by changing the detuning Δ to 'blue detuning', meaning that the excitonic fraction of the polaritons are increased relative to the photon component. Another alternative approach to EPs are indirect excitons in bilayer quantum wells systems[24]. Indirect excitons have also attracted much interest in recent years as candidates for the formation of BECs. This system has the advantage that the indirect exciton mass is much heavier than the EP mass, thus the conditions for achieving a Mott insulator state are more favorable. The disadvantage is that they are susceptible towards localization by disorder potentials in semiconductors, because of their relatively heavy mass.

For indirect excitons, another parameter d, which is the separation of two quantum wells, is available. The parameter d has the effect of controlling the Coulomb interaction between the electrons and holes. Figure 5 shows a typical semiconductor structure for (a) EPs and (b) indirect excitons. In both cases we assume that the particles experience periodic potentials by a suitable experimental configuration. Here we use the potential strength W_0 and d as parameters, and we calculate the on-site interaction and the tunneling between neighboring sites in the Hubbard models. For the calculation of interactions, we include the classical direct Coulomb interaction, exchange of holes, electrons, and excitons and the saturation effect due to the coupling of electrons and holes to the electromagnetic field. Figure 6 shows the main results. There units are normalized to $t_0 = h^2/8m_{pol}\lambda^2$, $U_0^{coul} = 2e^2 a_B |u|^4 / \pi^3 \varepsilon \lambda$, $U_0^{sat} = 2\hbar g (\pi/2)^{-0.5} |u|^2 \times \mathrm{Re}(uv^*) a_B^2/\lambda^2$, where h is the Plank constant, $\hbar g$ is the Rabi splitting of the EPs, m_{pol} is the EP effective mass, λ is the wave-length of periodic potential, e is the elementary charge, a_B is the Bohr radius of excitons, $|u|^2$ is the excitonic fraction of EPs, ε is the permittivity of GaAs. We see that increasing the potential strength W_0 decreases the hopping t while increasing U as expected. Increasing d enhances U^{coul} as the dipole moment of the excitons is enhanced with an increasing d.

In two dimensions, the phase transition is expected to occur at approximately $U/t \approx 23$. By increasing the potential W_0, it is clear that at some point U/t will reach this critical amplitude. For a potential size $W_0 \approx h^2/2m_{pol}\lambda$, the EP mass necessary to reach the phase transition is $m_{pol} \approx 10^{-3} \pi^3 h/e^2 a_B |u|^4 \approx 10^{-2} M$ where in the second relation we assumed that $|u|^2$ is if the order of unity and $a_B \approx 10$ nm, and M is the exciton mass. This corresponds to extremely exciton-like EPs, whose exciton fraction should be larger than 99% (see Fig. 2b).

Finally we consider the crossover from the polaritonic Mott insulator to the Fermi Mott insulators of electrons and holes in each layer by increasing the separation d. In the limit of infinite d, EPs cannot exist because the wave function of electrons and holes do not overlap, thus no coupling to photons are possible. In this limit, the Mott insulator should be formed respectively in each layer. A first order phase transition should occur between the BEC and Bose Mott insulator (BMI) of EPs, and between BCS phase of indirect excitons and Fermi Mott insulator of electrons and holes. BMI and FMI are expected to be smoothly connected in a similar way to a BEC-BCS crossover.

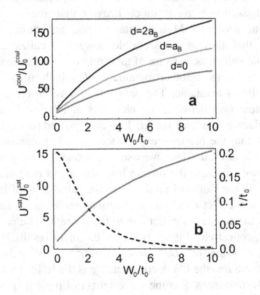

Figure 6. (from Ref. 24) (a) Calculated on-site interaction energy U^{coul}. (b) Tunneling term t (dashed line, right axis) and saturation interaction U^{sat} (solid line, left axis).

5. Accelerated Optimization Problem Search Using BECs

Quantum computation promises to offer great increases in speed over current computers due to the principle of superposition, where information can be processed in a massively parallel way[25]. The quantum indistinguishability[26] of particles, another fundamental principle of quantum mechanics, remains relatively unexplored in the context of information processing. Bosonic indistinguishability is the mechanism responsible for phenomena such as Bose-Einstein condensation[27] (BEC) and laser emission in the form of final state stimulation[28]. Here we show that by using bosonic particles it is possible to speed up the computation of a given NP-complete problem[29]. The method takes advantage of the fact that bosonic particles tend to concentrate in the minimal energy state at low temperatures. Since many computational problems, including NP-complete problems, can be reformulated as an energy minimization problem[30], the use of bosons allows the solution of the computational problem to be extracted with high probability. Furthermore, through the process of final state stimulation, the speed of attaining the solution is increased in comparison to the purely classical case. We find that an improvement in the speed of convergence to the ground state proportional to the number of bosons per site can be attained at a given error.

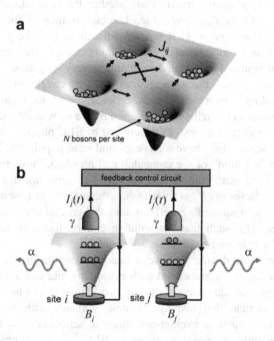

Figure 7. The schematic device configuration. (a) Each site of the Ising Hamiltonian is encoded as a trapping site, containing N bosons. The bosons can occupy one of two states $\sigma = \pm 1$, depicted as either red or blue. (b) The interaction between the sites may be externally induced by measuring the average spin on each site i via the detectors, which produce a detector current $I_i(t)$. The strength of the system-detector coupling is γ. A local field on each site equal to $B_i = \lambda \sum_j J_{ij} I_j(t)$ is applied via the feedback circuit. The system dissipates energy according to the coupling α to the environment.

5.1. The bosonic Ising model

We formulate the computational problem to be solved as an energy minimization problem of an Ising-type Hamiltonian[30]. For example, the NP-complete MAX-CUT problem, where the task is to group M vertices into two groups A and B such as to maximize the number of connections between the groups, is known to be equivalent to the Hamiltonian

$$H = \sum_{i,j=1}^{M} J_{ij} \sigma_i \sigma_j \qquad (8)$$

where J_{ij} is a real symmetric matrix that specifies the connections between the sites i, and $\sigma_i = \pm 1$ is a spin variable. The task is then to find the minimal energy spin configuration $\{\sigma_i\}$. It is very difficult to find the ground state of such a Hamiltonian by a brute force calculation because of the exponential (2^M) combinations of spins. In simulated annealing[31], very long annealing times are necessary to ensure that the system does not get caught in local minima. Quantum annealing[32] overcomes such problems due to local minima by introducing a quantum tunnelling term but requires a slow adiabatic evolution to prevent leaks into excited states. In the context of BEC physics, this scenario is very suggestive since BECs have a large ground state population, which can be regarded as the "solution" of the computational problem. Furthermore, through the process of final state stimulation, the speed of attaining this ground state is enhanced by a factor of $N + 1$ where N is the number of bosons. The aim of this work is to incorporate these effects into the ground state search problem specified by Eq. (8), such that the solution can be found with an enhanced probability and a reduced time overhead.

The computational device we have in mind is shown in Fig. 7. Each spin σ_i in the Hamiltonian (8) is associated with a trapping site containing N bosonic particles. The bosons can occupy one of two spin states, which we label by $\sigma = \pm 1$. Any particle that displays bosonic statistics with an internal spin state may be used, such as exciton-polaritons in semiconductor microcavities, which have recently observed to undergo BEC[17–20], or neutral atoms with an unpaired electron in atom chips[33]. Systems that undergo BEC are natural choices for implementation of such a device, since similar principles to the formation of a BEC are required in order for the rapid cooling to the solution of the computational problem. Exciton-polaritons possess a spin of $\sigma = \pm 1$ which can be injected by optical pumping with right or left circularly polarized laser beam. The sites are externally controlled such as to follow the Hamiltonian

$$H = \sum_{i,j=1}^{M} J_{ij} S_i S_j \qquad (9)$$

where $S_i = \sum_{k=1}^{N} \sigma_i^k$ is the total spin on each site i, and J_{ij} is the same matrix as in Eq. (8) which specifies the computational problem. As shown in Ref. 41, finding the ground state of (9) amounts to finding the ground state of (8), up to degeneracies involving frustrated spins. The interaction Hamiltonian may be induced by a system of measurement and feedback control (see Fig. 7b), where the total spin S_i on each site measured and fed back into the system by applying a local dc field on site i. Although in general J_{ij} has a large connectivity and is long-ranged, by using such a feedback method to induce the interactions there is no restriction to the kind of interactions J_{ij} that can be produced in principle (we refer the reader to Ref. 41 for further details).

Initially each site is prepared with equal populations of $\sigma = \pm 1$ spins, which can be achieved by using a linearly polarized laser, in the case of exciton-polaritons. The system is cooled in the presence of the interactions between the sites, by immersing the system in an external heat bath. The readout of the computation is simply performed by measuring the total spin on each site after the system cools down by dissipating heat into the environment. The sign of the total spin gives the information of $\sigma_i = \pm 1$ for the original spin model. Since the "computation" here is the cooling process itself, no complicated gate sequence needs to be employed to obtain the ground state.

5.2. Performance of the bosonic Ising model

To understand the effect of using bosons, first compare the thermal equilibrium configuration of a system described above with an equivalent system that uses classical, distinguishable particles. As a simple example, consider the two site Hamiltonian

$$H = -J S_1 S_2 - \lambda N (S_1 + S_2) \qquad (10)$$

where the second term is included such as there is a unique ground state due to the $S_i \leftrightarrow -S_i$ symmetry of the first term in the Hamiltonian. For a single spin on each site and $J, \lambda > 0$, the ground state configuration is $\sigma_1 = +1$, $\sigma_2 = +1$, which we regard as the "solution" of the computational problem. We neglect the presence of an on-site particle interaction $\propto S_i^2$ here since we assume that the strength of the interactions J produced by the induced feedback method can be made much larger than such a term which may occur naturally.

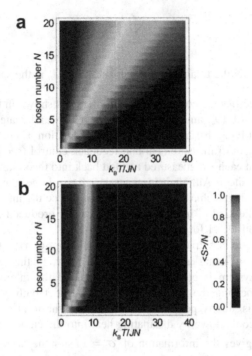

Figure 8. The average spin of the two site Ising Hamiltonian as a function of the boson number N and rescaled temperature $k_B T/JN$. (a) indistinguishable bosons and (b) classical distinguishable particles. The parameters used are $J = 10$ and $\lambda = 0.5$ in (10).

In Fig. 8 we show the average spin on a single site of the two site Hamiltonian, which can be calculated from evaluating the partition function assuming thermal equilibrium. Comparing bosonic particles and classical distinguishable particles, we see that the bosonic case has a larger average spin for $N > 1$ and all temperatures, corresponding to a spin configuration closer to the ground state. As the particle number is increased, the temperature required to reach a particular $\langle S_i \rangle$ increases. For the bosonic case, the temperature increases linearly with N, while for distinguishable particles it behaves as a constant for large N. This results in an improved signal to noise ratio for the bosons in comparison to distinguishable particles, which can systematically be improved by increasing N. The concentration of particles in the ground state configuration for bosons is precisely the same effect that is responsible for the formation of a BEC. Since the ground state corresponds to the solution of the computational problem, this corresponds to an enhanced probability of obtaining the correct answer at thermal equilibrium.

We now turn to the time taken to reach thermal equilibrium, after initially preparing the system with equal populations of $\sigma = \pm 1$ particles on each site. For simplicity consider first a single site version. We calculate the time dependence by evolving the rate equations including bosonic final state stimulation factors. Figure 9a shows the cooling of the system for $N = 1$ and $N = 50$ particles. As the number of particles is increased, we see that the equilibrium populations at $t \to \infty$ become dominated by the ground state of the system. We define the error probability ε as the probability of failing to obtain the correct ground state configuration after a single measurement of the total spin. Since we are only interested in whether the state is $\sigma = \pm 1$, only the sign of the measurement result is kept. By repeating the measurement many times it is possible to arbitrarily accurately estimate the correct sign of the spin.

In the device shown the Fig. 7, similar principles result in a speedup to the convergence of the ground state. We illustrate the accelerated cooling of the device by considering a typical four site Ising Hamiltonian (9) and numerically calculate the cooling starting from a $T = \infty$ configuration to obtain the expected time for a given error. We obtain results by extending Glauber's kinetic Ising model[34] to the bosonic case and using a kinetic Monte Carlo method. As is well known from past studies of simulated annealing, the presence of local minima slows down the time for equilibration dramatically. We choose an Ising problem that has a local minima in the spectrum between the states ↑↑↑↑ (local minimum) and ↑↑↓↓ (global minimum). The effect of the local minimum can be seen from the $N = 1$ case shown in Fig. 9b, where the time rapidly increases as $\varepsilon \to 0$ (i.e. $T \to 0$) unlike the single site case. As the boson number is increased, there is a significant speedup at constant error (Fig. 9bc). In general the speedups are greatest for low errors, due to a similar reason to the single site case. At lower errors we can expect even greater speedups due to a stronger final state stimulation effect. In Fig. 9c we plot the equilibration time as a function of $1/N$, where we obtain nearly linear behavior for large N. In all our numerical simulations we have found that bosons are able to speed up the equilibration times, although the exact degree of the speedup is dependent on the specific problem and the error and N chosen.

The origin of the speedup can be understood in the following simple way. The use of many bosons increases the energy scale of the Hamiltonian by a factor of $\sim N$. Due to bosonic statistics, the coupling of the spins to the environment is increased by a factor of $\sim N$. Thus by constructing a system out of bosons we have increased the energy scale of the entire problem by a factor of N, which results in a speedup of N. Spin flips due to random thermal fluctuations also occur on a timescale that is faster by a factor of N, resulting in a faster escape time out of local minima. We have confirmed this numerically in our simulations by measuring the average spin flip time τ^{flip}, defined as the average time for the total spin S_i on a particular site to change sign. The results in Fig. 9d show clearly a $1/N$ behavior, signifying that with an increased number of bosons, there are many more random spin flips per unit time. We can thus picture the system moving through the space of spin configurations with a speedup proportional to N. Since in a typical BEC there are a huge number of bosons, the speedup can be quite large in practice. For example, in atomic systems typically $N \approx 10^5$ or more bosons can be put in a single trap.

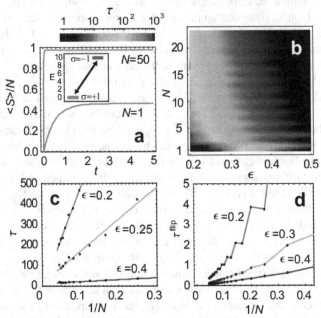

Figure 9. (a) Equilibration of a two level system with energy levels $(E_2 - E_1)/k_B T = 1$ and boson numbers as shown. The (b) equilibration time, (c) $1/N$ dependence of the equilibration time, and (d) average time between spin flips τ^{flip} for a 4 site bosonic Ising model with parameters $J_{ij} = -10, \lambda = -1$.

6. Summary and Conclusions

We have reviewed various approaches for quantum simulation with exciton-polariton BECs. From an experimental point of view, we have successfully implemented periodic lattice structures, and observed a new phase of condensation, namely excited state condensation. We have also investigated the possible scenario for phase transitions of Hubbard models with EPs. One aspect which we did not cover in this paper was the condensation of EPs in d-wave bands in two dimensional square lattices[43]. Such experimental and theoretical advances have revealed a rich variety of physics can be realized using this system.

BECs were also examined in the context of a device for solving optimization problems. We find that there is a linearly scaling improvement over implementing a standard Ising model, in terms of a decreased cooling time and reduced error of obtaining the correct ground state. We have tested our method on several problem specifications (i.e. J_{ij} matrices and site numbers), and in every case there are speedups that follow the qualitative behaviour of Fig. 9. The precise speedups that are obtainable are dependent on the particular problem chosen. Although the device discussed here is a computational device that uses quantum effects, it is rather different to a quantum computer in the standard sense, since the off-diagonal density matrix elements of the state of the device are explicitly zero at all times. Thus the device is closer to a classical computer that is speeded up using quantum effects, rather than a quantum computer. For these reasons the scaling of the equilibration time with the site number M should still be exponential[42], in analogy to the classical case. The speedup then manifests itself as a suppressed prefactor of this exponential function, which can be rather small as a result of the large number of bosons in a BEC. One advantage of the present scheme is that it is generally applicable to a large class of problems, including NP-complete problems, which can be formulated as a Hamiltonian to be optimized. Examining alternative models of quantum computation involving dissipative schemes may open up new possibilities towards the realization of a practical device.

Acknowledgments

This work is supported by the Special Coordination Funds for Promoting Science and Technology, the FIRST program for JSPS, Navy/SPAWAR Grant, N66001-09-1-2024, MEXT, and NICT.

References

1. I. Buluta and F. Nori, *Science* **326**, 108, (2009).
2. I. Bloch *et al.*, *Rev. Mod. Phys.* **80**, 885, (2008).
3. M. Greiner *et al.*, *Nature* **415**, 39, (2002).
4. R. Jördens *et al.*, *Nature* **455**, 204, (2008).
5. U. Schneider *et al.*, *Science* **322**, 1520, (2009).
6. R. Gerritsma *et al.*, *Nature* **463**, 68, (2010).
7. F. Zähringer *et al.*, *Phys. Rev. Lett.* **104**, 100503, (2010).
8. M. Imada *et al.*, *Rev. Mod. Phys.* **70**, 1039, (1998).
9. H. Tasaki, *Prog. Theor. Phys.* **99**, 489, (1998).
10. T. Moriya and K. Ueda, *Rep. Prog. Phys.* **66**, 1, (2003).

11. D. Jaksch et al., *Phys. Rev. Lett.* **81**, 3108, (1998).
12. T. Byrnes et al., *Phys. Rev. Lett.* **99**, 016405, (2006).
13. T. Byrnes et al., *Phys. Rev. B* **78**, 075320, (2008).
14. C. Weisbuch et al., *Phys. Rev. Lett.* 69, 3314, (1992).
15. H. Deng et al., *Rev. Mod. Phys.* **82**, 1489, (2010).
16. A. Imamoglu et al., *Phys. Rev. A* **53**, 4250, (1996).
17. H. Deng et al., *Science* **298**, 199, (2002).
18. H. Deng et al., *PNAS* **100**, 15318, (2003).
19. J. Kasprzak et al., *Nature* **443**, 409, (2006).
20. R. Balili et al., *Science* 316, 1007, (2007).
21. S. Utsunomiya et al., *Nature Physics* **4**, 700, (2008).
22. K. G. Lagoudakis et al., *Nature Physics* **4**, 706, (2008).
22. A. Amo et al., *Nature Physics* **5**, 805, (2009).
23. C. W. Lai et al., *Nature* **450**, 529, (2007).
24. T. Byrnes et al., *Phys. Rev. B* **81**, 205312, (2010).
25. M. Nielsen & I. Chuang, *Quantum Computation and Quantum Information* (Cambridge University Press, 2000).
26. C. Cohen-Tannoudji, B. Diu, F. Laloe, *Quantum Mechanics* (Wiley, 2006).
27. L. Pitaevskii & S. Stringari, *Bose-Einstein Condensation* (Oxford University Press, 2003).
28. W. Silfvast, *Laser Fundamentals* (Cambridge University Press, 2004).
29. G. Ausiello et al., *Complexity and Approximation* (Springer, 1999).
30. M. Mezard, G. Parisi, & M. Virasoro, *Spin Glass Theory and Beyond*, (World Scientific, 1987).
31. P. van Laarhoven & E. Aarts, *Simulated Annealing: Theory and Applications*, (D. Reidel Publishing Company, 1987).
32. A. Das & B. Chakrabarti, *Rev. Mod. Phys.* **80**, 001061 (2008).
33. Folman, R., Krueger, P., Schmiedmayer, J., Denschlag, J., Henkel, C. Microscopic atom optics: from wires to an atom chip. *Adv. At. Mol. Opt. Phys.* **48**, 263 (2002).
34. R. Glauber, *J. Math. Phys.* **4**, 294 (1963).
35. M. Swanson, *Path Integrals and Quantum Processes* (Academic Press, 1992).
36. H. Wiseman and G. Milburn, *Phys. Rev. Lett.* **70**, 548 (1993).
37. H. Wiseman, S. Mancini, J. Wang, *Phys. Rev. A* **66**, 013807 (2002).
38. H. Wiseman, G. Milburn, *Phys. Rev. A* **47**, 642 (1993).
39. D. Dalvit, J. Dziarmaga, W. Zurek, *Phys. Rev. Lett.* **86**, 373 (2001).
40. H. Hofmann, *J. Opt. B: Quantum Semiclass. Opt.* **7**, S208 (2005).
41. T. Byrnes, K. Yan, Y. Yamamoto, New J. Physics **13**, 113025 (2011).
42. K. Yan, T. Byrnes, Y. Yamamoto, Progress in Informatics **8**, 39 (2011).
43. N.Y. Kim et al., *Nature Phys.* **7**, 681 (2011).

QUANTUM SIMULATION USING ULTRACOLD ATOMS IN OPTICAL LATTICES

SEIJI SUGAWA[1,*], SHINTARO TAIE[1], REKISHU YAMAZAKI[1,2] and YOSHIRO TAKAHASHI[1,2]

[1]*Department of Physics, Graduate School of Science, Kyoto University, Kyoto-City, Kyoto, 606-8502 Japan*
[2]*JST-CREST, Chiyoda, Tokyo, 102-0075 Japan*
E-mail: ssugawa@scphys.kyoto-u.ac.jp

1. Introduction

In the past decades, the control of atoms using a laser field has made remarkable progress. The laser cooling technique for neutral atoms has been established, and has been applied to various studies. Especially, quantum degenerate dilute atomic gases has opened a new area of research. Bose-Einstein condensate (BEC) of dilute atomic gases realized in various atomic species[1-22] revealed the physical properties of weakly interacting Bose gas. Fermi degenerate gases was also created in various atomic species.[23-30]

Recent study using quantum degenerate gases has focused to the exploration of strongly-correlated system, which can be classified into two approaches. One approach is to use a Feshbach resonance[31] to tune the interparticle interaction from strong repulsion to strong attraction. Crossover from a BCS state, where fermion form cooper pairing, to a molecular BEC was successfully realized in a Fermi gas using this method.[32] The property of unitary Fermi gas and strongly-interacting Bose gas are extensively studied near the Feshbach resonance.

Another approach is to load atoms into a periodical potential created by laser light, which is called an optical lattice. Atoms confined in the lowest Bloch band of the optical lattice is described by a single-band Hubbard model, which was first pointed out by Jaksch et al.[33] Although a Hubbard model only takes into account on-site interaction and nearest-neighbor hopping, it can explain various phenomena, such as anisotropic superconductivity and itinerant magnetism, where electronic correlation plays an important role. The experimental milestone was the observation of superfluid to Mott insulator transition of bosonic atom.[34] Subsequently, band insulator, and metal to Mott phase transition in a three-dimensional cubic lattice was realized in a two-component fermionic system.[35–37]

1.1. *Quantum simulation of Hubbard model*

An optical lattice system is a new quantum many-body system. Atoms confined in an optical lattice has essentially no defects or impurity. The high controllability of the system is also a large advantage. Important parameters in the Hubbard model can be controlled independently. One can tune the on-site interaction by Feshbach resonance and the ratio of on-site interaction over hopping energy by changing the lattice laser intensity. The temperature or the entropy of the system can be changed by initial temperature before loading into the optical lattice. The average filling can also be changed by the atom number or the confinement potential. One can therefore experimentally simulate the Hubbard model and investigate phase diagrams. Following Richard. P. Feynman,[38] it is often referred to as quantum simulation of Hubbard model. Here, quantum simulation refers to simulate a quantum system by another controllable quantum system.

1.2. *Why quantum simulation?*

The quantum simulation is important because it may give answers to unsolved questions in condensed matter systems. One of the most important challenge in quantum simulation of Hubbard model is to simulate two-dimensional repulsive Fermi Hubbard model to get insight into the origin of d-wave superconductivity in cuprate superconductor. There is no common understanding on the behaviors of d-wave superconductivity in cuprate superconductor yet. Theoretical calculations have shown that d-wave superconductivity is expected to appear near the hole doping region away from half-filling in two-dimensional repulsive Fermi Hubbard model. However, the calculation is based on approximation and there is no guarantee for the result. Unfortunately, it is difficult to "solve" the Hubbard model exactly in a large system size. If one can solve the Hubbard model experimentally in an optical lattice, we may better understand two-dimensional repulsive Fermi Hubbard system. It is also possible to check whether the approximation used in theory is valid or not.

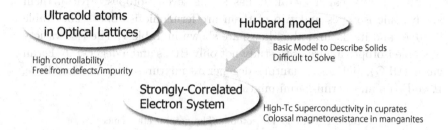

Fig. 1. Concept of quantum simulation of Hubbard model.

1.3. Extending the system

One can also further extend the model easily by dealing with more complex system, including a system that has no counterpart in real material. By loading two-species mixtures in an optical lattice, a Bose-Fermi Hubbard model, a Fermi-Fermi Hubbard model or a Bose-Bose Hubbard model can be implemented. Various novel phases are expected to appear at low temperature. One can also deal with multicomponent fermion with more than three spins. Theoretical studies show that the behavior is different from the case of two-component system. For example, in anisotropic interaction, color superfluidity are expected to appear.[39] Enlarged spin symmetry can be explored with nuclear spins of fermionic isotope of two-electron system such as ^{173}Yb ($I=5/2$) or ^{87}Sr ($I=9/2$).[40]

Various lattice geometries such as triangular lattice or a Kagomé lattice are realized to explore frustrated system.[41] A superlattice can be also implemented. Atomic species with large magnetic dipole moment such as chromium, dysprosium and erbium, or polar molecule with large electric dipole moment can be used to explore the effect of the long-range, non-local, anisotropic interaction. One can also add disorder by speckle light or impurity atoms. As recently demonstrated, spin-orbit coupling can be also implemented.[42,43] Extending the model step by step to a more complex system is an important step to gain a new insight into the quantum many-body system by quantum simulation.

2. An approach using ytterbium

While most of ultracold-atom experiments use alkali-metal atoms such as rubidium, potassium and lithium, we are using ytterbium (Yb) atoms to perform quantum simulation. Here we review the unique features of Yb for ultracold-atom experiments.

Yb is a rare earth metal. It has stable seven isotopes. Five of them are bosonic isotopes and two of them are fermionic isotopes. The stable isotopes and its natural abundance are shown in Table 1. The rich variety of stable isotopes allows us to study not only BECs and a degenerate Fermi gases (DFG), but also quantum degenerate mixtures such as Bose-Bose, Bose-Fermi and Fermi-Fermi mixtures.

Table 1. Stable Yb isotopes. The natural abundance and the nuclear spin of each isotope are also shown.

Atomic Mass	Natural abundance	Nuclear spin
168	0.13%	0
170	3.05%	0
171	14.3%	1/2
172	21.9%	0
173	16.12%	5/2
174	31.8%	0
176	12.7%	0

The important parameters to characterize ultracold atomic gases are the s-wave scattering lengths. The scattering lengths govern the efficiency of the evaporative cooling and properties of realized quantum phases. It is also necessary to determine the on-site interaction of the Hubbard model. A photoassociation (PA) experiment[44] has precisely determined the scattering lengths of all the combinations of Yb isotopes using mass-scaling law. The s-wave scattering lengths of all combination of Yb is shown in Table 2. Note that the two-body interaction at low temperature is given by

$$U(\boldsymbol{r}) = g\delta(\boldsymbol{r}) = \frac{4\pi\hbar^2 a}{m}\delta(\boldsymbol{r}), \qquad (1)$$

where a is the s-wave scattering length and m is the atomic mass.

The ground state electronic configuration is [Xe] $4f^{14}\,6s^2$. Due to the two-valence electrons, it shares similar properties with that of alkaline-earth metal. The energy levels are shown in Fig. 2. The ground state of Yb is $^1S_0$ state. Although there is no electron spin in the ground state, fermionic isotopes have nuclear spins. Especially, the six spin components of ^{173}Yb with the nuclear spin $I=5/2$ enables us to explore the enlarged spin symmetry. The enlarged spin symmetry is realized by the fact that the two-body interaction does not depend on the spin component. From a quantum information perspective, as the nuclear spins, decoupled from the electronic state, are robust against decoherence, ^{171}Yb is also a promising candidate for qubits.[45,46]

Table 2. s-wave scattering lengths for all Yb isotopic combinations determined by two-photon photoassociation (PA). The units are in nm. The underlines show that the corresponding isotopic combinations has been cooled down to quantum degenerate regime.

	^{168}Yb	^{170}Yb	^{171}Yb	^{172}Yb	^{173}Yb	^{174}Yb	^{176}Yb
^{168}Yb	13.33(18)						
^{170}Yb	6.19(8)	3.38(11)					
^{171}Yb	4.72(9)	1.93(13)	-0.15(19)				
^{172}Yb	3.44(10)	-0.11(19)	-4.46(36)	-31.7(3.4)			
^{173}Yb	2.04(13)	-4.30(36)	-30.6(3.2)	22.1(7)	10.55(11)		
^{174}Yb	0.13(18)	-27.4(2.7)	22.7(7)	10.61(12)	7.34(8)	5.55(8)	
^{176}Yb	-19.0(1.6)	11.08(12)	7.49(8)	5.62(8)	4.22(10)	2.88(12)	-1.28(23)

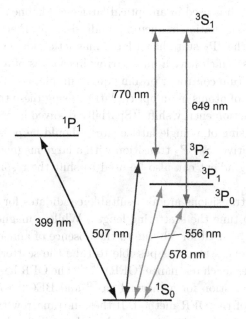

Fig. 2. Energy diagram of Yb.

For the laser cooling of Yb, there are two important optical transitions. One is the $^1S_0$-$^1P_1$ transition, which has a natural linewidth of 29 MHz. This strong dipole-allowed transition is applied for Zeeman slower and absorption imaging. The other transition is the $^1S_0$-$^3P_1$ intercombination transition, which has a natural linewidth of 182 kHz. This intercombination transition is applied to Magneto-Optical Trap (MOT), where the Doppler cooling limit is 4.4 μK. The properties of two optical transition are shown in Table 3.

Table 3. Properties of optical transitions of $^1S_0$-$^1P_1$ and $^1S_0$-$^3P_1$.

Optical transition	$^1S_0$-$^1P_1$	$^1S_0$-$^3P_1$
Wavelength [nm]	398.9	555.8
Linewidth [MHz]	29.1	0.182
Lifetime [ns]	5.46	875
Doppler cooling limit [μK]	690	4.4
Saturation Intensity [mW/cm^2]	60	0.14

There are also two useful ultranarrow transitions in Yb. One is the $^1S_0$-$^3P_0$ transition, which is used in an optical lattice clock, one of the primary candidates of next-generation frequency standards. The other is the $^1S_0$-$^3P_2$ transition, where the $^3P_2$ state has a large magnetic moment. Both ultranarrow transitions, which have a ultranarrow linewidths of about 10 mHz, are useful for quantum control or probing quantum phases. High-resolution laser spectroscopy of atoms in an optical lattice using these transitions can probe small interaction energy shift.[47] Spatially-resolved laser spectroscopy or spectral addressing of a single lattice site[45] should be possible by using the magnetic-sensitive $^1S_0$-$^3P_2$ transition with a large magnetic field gradient. A light-shift gradient can also be used to shift the resonance of both transitions.

These optical transitions are also suitable candidates for optical Feshbach resonance to tune the scattering length. While a magnetic Feshbach resonance cannot be used for Yb due to the absence of an electron spin in the electronic ground state, it is possible to tune the scattering length by using an optical Feshbach resonance (OFR).[48,49] The OFR has been applied to the $^1S_0$-$^3P_1$ transition for a thermal gas[50] and BEC[51] to demonstrate the powerfulness of the OFR method. If these ultranarrow transitions are used, large tuning of the scattering length with negligible heating will be realized.

3. Production of quantum degenerate Yb atoms

All-optical formation of quantum degenerate Yb atoms has been successfully performed for various isotopes and their mixtures. The experimental procedure is as follows. First, Yb atomic beam produced from an atomic oven is decelerated by a Zeeman slower laser beam tuned near the $^1S_0$-$^1P_1$

transition (natural linewidth ~28 MHz). Atoms are then collected by a MOT with the MOT laser beam tuned near the $^1S_0$-$^3P_1$ transition (natural linewidth ~182 kHz). A mixture of two Yb isotopes is produced with two-color MOT laser beams, which have a frequency difference close to an isotope shift, with the laser frequency of Zeeman slower beam being swept from one to another to load both isotopes. After the loading, the MOT is compressed, and the atoms are further cooled for transfer to a crossed far-off-resonant trap (FORT). In the crossed FORT, the potential depth is decreased to perform forced evaporation until a quantum degenerate region is reached.

So far, various kinds of quantum degenerate gases and mixture of Yb isotopes have been created.[11–14,27,28] We have been able to produce a BEC of ^{168}Yb, ^{170}Yb, and ^{174}Yb, and a DFG of ^{171}Yb and ^{173}Yb. As for Yb mixtures, a ^{170}Yb-^{173}Yb and a ^{173}Yb-^{174}Yb Bose-Fermi mixtures, a ^{174}Yb-^{176}Yb and a ^{168}Yb-^{174}Yb Bose-Bose mixtures, and a ^{171}Yb-^{173}Yb Fermi-Fermi mixture with SU(2) × SU(6) spin symmetry has been also successfully produced. The availability to create a various kind of quantum degenerate gases and mixtures of Yb surely enhance the accessible problem in quantum simulation of Hubbard model. Time-Of-Flight (TOF) images of a ^{174}Yb-^{173}Yb Bose-Fermi mixture are shown in Fig. 3.

4. Superfluid-Mott insulator transition

Using TOF absorption imaging technique, phase coherence of atoms in an optical lattice can be measured. Here we show probing of superfluid to Mott insulator transition by observing the matterwave interference patterns in TOF absorption images.

When the interaction can be neglected, the density distribution after TOF of time with t can be expressed as[52]

$$n(\boldsymbol{r},t) = \left(\frac{m}{\hbar t}\right)^3 \left|\tilde{w}(\boldsymbol{k}=\frac{m\boldsymbol{r}}{\hbar t})\right|^2 S(\boldsymbol{k}=\frac{m\boldsymbol{r}}{\hbar t}), \qquad (2)$$

where $b_i^\dagger(b_i)$ is the creation (annihilation) operator at lattice site i, $\tilde{w}$ is the Fourier transform of the Wannier function, which determines the envelope of the density distribution, and

$$S(\boldsymbol{k}) = \sum_{i,j} e^{i\boldsymbol{k}\cdot(\boldsymbol{r}_i-\boldsymbol{r}_j)} \langle b_i^\dagger b_j \rangle, \qquad (3)$$

describes the interference pattern. When long-range phase coherence exists, a correlation function $\langle b_i^\dagger b_j \rangle$ varies slowly across the lattice and thus $S(\boldsymbol{k})$ shows a shape interference pattern.

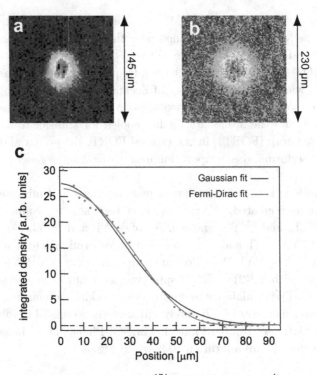

Fig. 3. Simultaneous TOF imaging of (a) ^{174}Yb BEC ($N_B = 1 \times 10^4$) with no discernible thermal component and (b) ^{173}Yb DFG ($N_F = 2.0 \times 10^4$) in a quantum degenerate ^{174}Yb-^{173}Yb Bose-Fermi mixture. The fermionic isotope of ^{173}Yb is prepared in an almost equal population of six nuclear spin components. The images are taken at TOFs of 14 ms and 15 ms, respectively in a single experimental run. An asymmetric expansion of BEC can be seen. The temperature is evaluated to be $T = 30 nK$ and $T/T_F = 0.17$ from the 2D Fermi-Dirac fit from the density profile of a ^{173}Yb. (c) Azimuthal averaged density profile of ^{173}Yb. The density profile fit well with the Fermi-Dirac fit (red curve). The Gaussian fit to the density profile is also shown (blue curve). The deviation from a Gaussian fit is clearly seen in the wing of the cloud.

As expected from the Bose-Hubbard model, the system is in a superfluid phase when the lattice depth is shallow and the dimensionless interaction strength (U/J) is below the critical value. The superfluid phase is characterized by the existence of long-range phase coherence. Therefore clear matterwave interference should be observed in the TOF image. However, when the lattice depth is increased and exceed the critical value, the system enters into the Mott insulating phase and long-range coherence is lost. Thus in the Mott phase, a clear interference pattern should diminish.

An illustrating example of TOF absorption images of matterwave interference patterns for bosonic isotope of ^{168}Yb is shown in Fig. 4. In the experiment, ^{168}Yb BEC, created in an optical trap, is adiabatically loaded into a three-dimensional optical lattice. After loaded into the corresponding lattice depth, all the trap potential is abruptly switched off to let the atoms freely expand. When the lattice depth is lower than the Mott transition point of 8.5E_R for unit filling, a clear interference pattern is observed (upper column in Fig. 4). However, when the final lattice depth exceed 9E_R, an interference pattern begins to disappear in increasing the lattice depth. To verify that the disappearance of interference pattern is not due to the heating effect, the lattice depth is ramped back to a shallow lattice depth of 5E_R, where we observe a clear matterwave interference again, after loading into 12E_R where phase coherence is lost.

The above explained behavior can be clearly seen in the central width of interference pattern. Figure 5 shows the center peak width as a function of lattice depth obtained from analyzing the TOF images in Fig. 4. Below the Mott transition point, the central peaks show small dependence on lattice depth, followed by an abrupt kink around the Mott transition point.

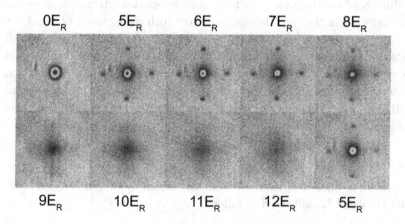

Fig. 4. Absorption imaging of matterwave interference patterns of ^{168}Yb. At shallow lattice depth, interference patterns with first-order peaks at $\Delta x = (\hbar k/m)t$ is observed (upper column). When entering the Mott transition point of 8.5E_R for unit filling, clear interference patterns are diminished (lower column). To verify that the disappearance of interference pattern is not due to the heating effect, the lattice depth is ramped back to 5E_R, after loading into 12E_R (bottom right).

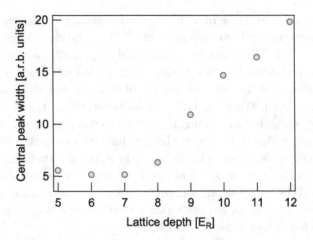

Fig. 5. Center peak width as a function of lattice depth obtained from TOF images in Fig. 4.

5. Strongly-correlated phases in Bose-Fermi mixtures

Here, we show an exotic system realized with bosonic and spinful fermionic ytterbium isotope. It is easy to expect that a drastic change will occur when two different Mott insulators with strong interspecies interaction are combined together in the same lattice. We show such competition of two Mott insulating phases results in diverse behavior. The schematic illustration of effects of interspecies interaction and filling on such dual Mott insulators of bosons and fermions is shown in Fig. 6. As shown in the figure, in addition to interspecies interactions, the relative fillings of two species produce an additional kind of diversity to this system, which provides us with a novel paradigm, such as filling-induced phases.[53]

5.1. *Hamiltonian of the system*

The system of dual Mott insulator is characterized by the single-band Bose-Fermi Hubbard Hamiltonian as follows,

$$\hat{H}_{\mathrm{BF}} = \hat{H}_{\mathrm{B}} + \hat{H}_{\mathrm{F}} + U_{\mathrm{BF}} \sum_{i,\sigma} \hat{n}_{\mathrm{B},i}\hat{n}_{\mathrm{F},i,\sigma}, \tag{4}$$

$$\hat{H}_{\mathrm{B}} = -J \sum_{\langle i,j \rangle} \hat{b}_i^\dagger \hat{b}_j + \frac{U_{\mathrm{BB}}}{2} \sum_i \hat{n}_{\mathrm{B},i}(\hat{n}_{\mathrm{B},i} - 1) + \sum_i \epsilon_i \hat{n}_{\mathrm{B},i}, \tag{5}$$

$$\hat{H}_{\mathrm{F}} = -J \sum_{\langle i,j \rangle,\sigma} \hat{f}_{i,\sigma}^\dagger \hat{f}_{j,\sigma} + \frac{U_{\mathrm{FF}}}{2} \sum_{i,\sigma \neq \sigma'} \hat{n}_{\mathrm{F},i,\sigma}\hat{n}_{\mathrm{F},i,\sigma'} + \sum_{i,\sigma} \epsilon_i \hat{n}_{\mathrm{F},i,\sigma}, \tag{6}$$

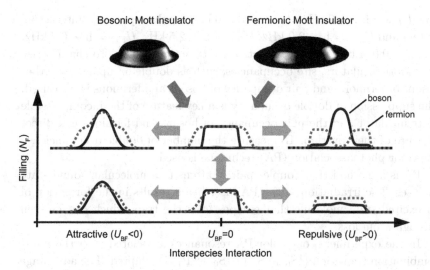

Fig. 6. The effects of interspecies interaction and filling on dual Mott insulators of bosons and fermions is schematically illustrated. The density distributions of each Mott insulator of unit occupancy is strongly altered by the interspecies interaction and the filling.

where $\langle i,j \rangle$ denotes summation over nearest-neighbors and $\sigma \in \{1,2,...,6\}$ denotes the spin component of fermionic atoms. $\hat{b}_i^\dagger$ ($\hat{f}_{i,\sigma}^\dagger$) and $\hat{b}_i$ ($\hat{f}_{i,\sigma}$) are creation and annihilation operators for bosons (fermions with σ spin). $\hat{n}_{B,i} = \hat{b}_i^\dagger \hat{b}_i$ ($\hat{n}_{F,i,\sigma} = \hat{f}_{i,\sigma}^\dagger \hat{f}_{i,\sigma}$) are the number operators for bosons (single spin component fermions). U_{BB}, U_{FF}, and U_{BF} represent on-site interaction energies between bosons, fermions with different spin components, and boson-fermion pairs, respectively. Here $U_{BB} < |U_{BF}| < U_{FF}$ is satisfied in both systems. J is the tunneling matrix element, and ϵ_i is the energy offset due to an external confinement potential.

The dual Mott insulator system is prepared as follows. First, a quantum degenerate Bose-Fermi mixture of repulsively interacting ^{174}Yb-^{173}Yb or attractively interacting ^{170}Yb-^{173}Yb is created in an optical trap produced by crossed laser beams with a wavelength of 532 nm by sympathetic evaporative cooling of bosonic and fermionic species. After preparing the quantum degenerate mixture in the optical dipole trap, the atoms are loaded into a three dimensional optical lattice with a lattice constant of 266 nm in a simple cubic symmetry. The final lattice depth is set to 15 E_R for repulsive system and 18 E_R for attractive system. At the final lattice depth, the Hubbard parameter is calculated to be $U_{BB} = h \times 2.6$ kHz, $U_{BF} = h \times 3.5$

kHz, $U_{FF} = h \times 5.0$ kHz, $J = h \times 0.026$ kHz for the repulsively interacting system and $U_{BB} = h \times 2.0$ kHz, $U_{BF} = h \times -2.5$ kHz, $U_{FF} = h \times 6.2$ kHz, $J = h \times 0.014$ kHz for the attractively interacting system. To characterize dual Mott insulators, site occupancies, such as double occupancy of either bosons or fermions and pair occupancy of bosons and fermions is measured. The suppression of double occupancy is a key feature of the incompressible Mott insulator and the pair occupancy of bosons and fermions is a direct measure of their overlap. To measure the number of double occupancies, a one-color photoassociation (PA) technique is used.

PA is a method that couples pair of atoms to a molecular bound state by laser. The irradiation of the PA laser beam results in the formation of an excited state molecule that escapes from the trap, which can be seen through atom loss.

In the experiment, one-color PA resonances associated with the intercombination transition ($^1S_0 + ^1S_0 \leftrightarrow ^1S_0 + ^3P_1$) is adapted. The advantage of PA method is that, in addition to the isotope selectivity, several types of occupancy can be measured in the same system by simply changing the laser frequency, as many PA resonances exist for any kind of atom pair in general.[54] Also, the occupancy measurement can be performed in a short time by choosing appropriate PA resonance and laser intensity.

Before the occupancy measurement, to freeze out the site occupancy during the PA process, the lattice depth is further ramped up to 25 E_R in a short time of 2 ms, which is at the same time long enough to suppress the non-adiabatic excitation to higher bands. During the measurement, a strong enough PA laser beam is applied to deplete all the doubly occupied sites. The occupancy measurement is performed by changing the number of fermionic atoms (N_F), in repulsively and attractively interacting dual Mott insulator system, repulsively.

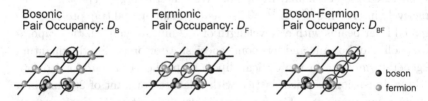

Fig. 7. Three types of pair occupancy are measured in an optical lattice. The atoms circled are depleted by photoassociation in the respective measurement.

5.2. Repulsively interacting Bose-Fermi system

In the repulsively interacting Bose-Fermi system ($U_{BF} > 0$), the system is found to be characterized by two distinct phases depending on N_F. When N_F is small, bosonic, fermionic and boson-fermion pair occupancies is found to take values near zero. Numerical calculation show that $n_B + n_F = 1$ is satisfied in this regime. We thus identify this phase as a mixed Mott state of bosons and fermions, where a total set of bosons and fermions satisfies commensurability, but each species separately does not. As N_F increases, bosonic double occupancy monotonically increases. However, fermion pair occupancy and bose-fermi pair occupancy still takes a value near zero, even at larger numbers of fermions. This behavior is identified as a phase separation of boson and fermion from the numerical calculation. The behavior of pair occupancies can be understood from the relative strength of the on-site interaction parameters ($U_{BB} < U_{FF} < U_{BF}$). As the fermion on-site interaction U_{FF} is the largest, it is energetically favorable for fermions to spatially extend to avoid double occupancy. As a result of the Bose-Fermi repulsive interaction U_{BF}, bosons and fermions repel each other, which results in a phase-separation.

A key feature of Mott insulator is the existence of the energy gap known as the Mott gap in the excitation spectrum. By modulating the lattice depth as $V(t) = V_0 + dV\sin(\omega t)$ at the angular frequency of ω, excitation gap can be measured. A schematic illustration of the lattice modulation spectroscopy is shown. At the modulation frequency corresponding to the on-site interaction, particle-hole excitation occurs in the Mott insulator.

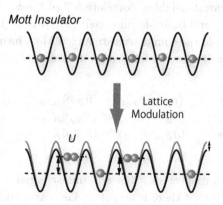

Fig. 8. Schematic illustration of modulation spectroscopy.

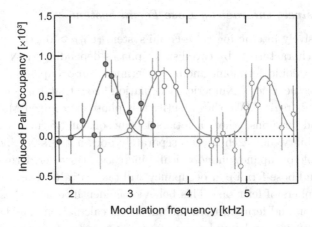

Fig. 9. Modulation spectroscopy of mixed Mott insulator. Bosonic pair occupancy (closed circle), fermionic pair occupancy (dotted circle) and Bose-Fermi pair occupancy (open circle) are measured, after a sinusoidal oscillation of lattice depth. The three peaks correspond to the energy gaps with on-site interaction energy of U_{BB}, U_{BF}, and U_{FF}, respectively.

The resonance can be determined by pair/double occupancy measurement or heating due to the absorbed energy in the case of boson.

Besides the pair occupancy measurement, we performed the modulation spectroscopy[55] in the mixed Mott regime (see Fig. 9). The three peaks corresponding to the Mott gaps with on-site interaction energy (U_{BB}, U_{FF} and U_{BF}) have been be clearly observed. When the effect of the lattice modulation can be treated in a linear response regime, the relative peak weight reflects the nearest-neighbor correlation[56] of bosons ($\langle \hat{n}_{B,i} \hat{n}_{B,i+1} \rangle$), fermions ($\langle \hat{n}_{F,i} \hat{n}_{F,i+1} \rangle$) and boson-fermion pair ($\langle \hat{n}_{B,i} \hat{n}_{F,i+1} \rangle$). For example, depending on the type of the commensurate mixed Mott insulator, nearest-neighbor correlation will take the following relation.

Paramagnetic phase: $\langle \hat{n}_{B,i} \hat{n}_{F,i+1} \rangle \sim \langle \hat{n}_{B,i} \hat{n}_{B,i+1} \rangle \sim \langle \hat{n}_{F,i} \hat{n}_{F,i+1} \rangle$

Phase separated phase: $\langle \hat{n}_{B,i} \hat{n}_{F,i+1} \rangle \ll \langle \hat{n}_{B,i} \hat{n}_{B,i+1} \rangle \sim \langle \hat{n}_{F,i} \hat{n}_{F,i+1} \rangle$

Charge density wave: $\langle \hat{n}_{B,i} \hat{n}_{F,i+1} \rangle \gg \langle \hat{n}_{B,i} \hat{n}_{B,i+1} \rangle \sim \langle \hat{n}_{F,i} \hat{n}_{F,i+1} \rangle$

The observed excitation spectrum shows a almost equal weights of three peaks, which manifests that there is no particular nearest-neighbor correlation. If the system is phase separated, the peak at the frequency of U_{BF}/h should be strongly suppressed. Together with the measured suppression of all the three pair occupancies, the mixed Mott state with paramagnetic order is shown to be realized.

5.3. *Attractively interacting Bose-Fermi system*

In the attractively interacting Bose-Fermi system ($U_{BF} < 0$), a large difference between the repulsively and attractively interacting systems is observed, which highlights the important role of interspecies interaction in the strongly interacting dual Mott insulators. The clear difference from the behavior of the repulsively interacting system is seen in the boson-fermion pair occupancy measurement, where the significantly large pair occupancy is observed. The measured bosonic pair occupancy shows a non-monotonic dependence on N_F values. We identify three phases. Each phase is found to be characterized by a formation of different types of on-site composite particles near the trap center. The transition from one phase to another can be understood qualitatively by considering the relative strength of the various types of on-site interactions and the associated composite particles.

5.4. *Thermodynamics*

The thermodynamics of the dual Mott insulators was also revealed. Using the numerically calculated bosonic double occupancy as a thermometry in optical lattices, the temperature realized experimentally in the lattice is evaluated.

Depending on the interspecies interaction, two distinct behaviors were confirmed. In the repulsive case, the system was found to adiabatically cooled during the lattice loading. This was due to the fact that the large spin degrees of freedom, which amounts to 7, helped carrying large spin entropy to cool the system, similar to Pomeranchuk effect in ^{3}He.[57]

On the other hand, in the attractively interacting case, the creation of composite particle within the external confinement trap results in reduction of number of particles involved in the system, and thus the system was found to be adiabatically heated.[58,59]

6. Prospect

The present work also paves the way for exploring diverse quantum phases of mixtures around the Mott transitions at lower temperatures such as the supersolid phase, alternating Mott insulator[60] in the repulsively interacting Bose-Fermi mixture and the novel phases by composite particles in the attractively interacting Bose-Fermi mixture.[61] At sufficiently low temperatures, the quantum statistics of atoms will play an important role in these realized phases.

The present work will also give insights into performing quantum simulation of fermionic Hubbard model. Exploring the hole doping region is important in the quantum simulation of Hubbard model. However, due to the inhomogeneous trapping potential, a well-controlled hole doping may be difficult. This is because the inhomogeneous trapping leads to inhomogeneous density distribution of atoms, and thus the hole will not be doped homogeneously. Nonetheless, as recently proposed, well-controlled hole doping will be possible using the system of mixed Mott insulator, which can be mapped to the hole doped homogeneous t-J model[62].

To perform quantum simulation of quantum magnetism, it is the utmost importance to realize a temperature of the superexchange spin-spin coupling. Up to now, quantum magnetism in an optical lattice has not realized yet. Various techniques have been proposed, such as to manage the trap potential to squeeze out the entropy to the outer region[63] or to dynamically change the lattice potential to convert the band insulator to antiferromagnetic ordering.[64] The present work will also give insights into this problem. Atomic mixture may also be used to cool another isotope by entropy transfer during the lattice loading. Entropy transfer in a atomic mixture has been realized in a species selective dipole trap.[65] In an optical lattice, the entropy can be transfered to spin entropy of localized fermion. As we have seen in the repulsive dual Mott insulator, the large spin degrees of freedom of ^{173}Yb helped cooling the system in the phase separated state. The fermionic isotope of ^{173}Yb is especially promising for this purpose as it can store large entropy of $k_B \ln(6)$ per site.[66]

Acknowledgement

The work on strongly-correlated phases in Bose-Fermi mixtures is a collaborative work with Dr. Kensuke Inaba and Dr. Makoto Yamashita at NTT basic research laboratory. This work is supported by the Grant-in-Aid for Scientific Research of JSPS (Nos.18204035, 21102005C01 (Quantum Cybernetics)), GCOE Program "The Next Generation of Physics, Spun from Universality and Emergence" from MEXT of Japan, and World-Leading Innovative R&D on Science and Technology (FIRST).

References

1. M. H. Anderson, J. R. Ensher, M. R. Matthews, C. E. Wieman and E. A. Cornell, *Science* **269**, 198 (1995).

2. K. Davis, M. Mewes, M. Andrews, N. Van Druten, D. Durfee, D. Kurn and W. Ketterle, *Physical Review Letters* **75**, 3969 (1995).
3. S. L. Cornish, N. R. Claussen, J. L. Roberts, E. A. Cornell and C. E. Wieman, *Phys. Rev. Lett.* **85**, 1795 (2000).
4. G. Roati, M. Zaccanti, C. D'Errico, J. Catani, M. Modugno, A. Simoni, M. Inguscio and G. Modugno, *Phys. Rev. Lett.* **99**, p. 010403 (2007).
5. G. Modugno, G. Ferrari, G. Roati, R. J. Brecha, A. Simoni and M. Inguscio, *Science* **294**, 1320 (2001).
6. C. C. Bradley, C. A. Sackett, J. J. Tollett and R. G. Hulet, *Phys. Rev. Lett.* **79**, p. 1170 (1997).
7. K. B. Davis, M. O. Mewes, M. R. Andrews, N. J. van Druten, D. S. Durfee, D. M. Kurn and W. Ketterle, *Phys. Rev. Lett.* **75**, 3969 (1995).
8. T. Weber, J. Herbig, M. Mark, H.-C. Nagerl and R. Grimm, *Science* **299**, 232 (2003).
9. A. Robert, O. Sirjean, A. Browaeys, J. Poupard, S. Nowak, D. Boiron, C. I. Westbrook and A. Aspect, *Science* **292**, 461 (2001).
10. F. Pereira Dos Santos, J. Léonard, J. Wang, C. J. Barrelet, F. Perales, E. Rasel, C. S. Unnikrishnan, M. Leduc and C. Cohen-Tannoudji, *Phys. Rev. Lett.* **86**, 3459 (2001).
11. S. Sugawa, R. Yamazaki, S. Taie and Y. Takahashi, *Phys. Rev. A* **84**, p. 011610 (2011).
12. Y. Takasu, K. Maki, K. Komori, T. Takano, K. Honda, M. Kumakura, T. Yabuzaki and Y. Takahashi, *Phys. Rev. Lett.* **91**, p. 040404 (2003).
13. T. Fukuhara, S. Sugawa and Y. Takahashi, *Phys. Rev. A* **76**, p. 051604 (2007).
14. T. Fukuhara, S. Sugawa, Y. Takasu and Y. Takahashi, *Phys. Rev. A* **79**, p. 021601 (2009).
15. S. Kraft, F. Vogt, O. Appel, F. Riehle and U. Sterr, *Phys. Rev. Lett.* **103**, p. 130401 (2009).
16. S. Stellmer, M. K. Tey, B. Huang, R. Grimm and F. Schreck, *Phys. Rev. Lett.* **103**, p. 200401 (2009).
17. Y. N. M. de Escobar, P. G. Mickelson, M. Yan, B. J. DeSalvo, S. B. Nagel and T. C. Killian, *Phys. Rev. Lett.* **103**, p. 200402 (2009).
18. S. Stellmer, M. K. Tey, R. Grimm and F. Schreck, *Phys. Rev. A* **82**, p. 041602 (2010).
19. P. G. Mickelson, Y. N. Martinez de Escobar, M. Yan, B. J. DeSalvo and T. C. Killian, *Phys. Rev. A* **81**, p. 051601 (2010).
20. D. G. Fried, T. C. Killian, L. Willmann, D. Landhuis, S. C. Moss, D. Kleppner and T. J. Greytak, *Phys. Rev. Lett.* **81**, 3811 (1998).
21. A. Griesmaier, J. Werner, S. Hensler, J. Stuhler and T. Pfau, *Phys. Rev. Lett.* **94**, p. 160401 (2005).
22. M. Lu, N. Q. Burdick, S. H. Youn and B. L. Lev, *Phys. Rev. Lett.* **107**, p. 190401 (2011).
23. B. DeMarco and D. Jin, *Science* **285**, p. 1703 (1999).
24. A. Truscott, K. Strecker, W. McAlexander, G. Partridge and R. Hulet, *Science* **291**, p. 2570 (2001).
25. F. Schreck, L. Khaykovich, K. L. Corwin, G. Ferrari, T. Bourdel, J. Cubizolles and C. Salomon, *Phys. Rev. Lett.* **87**, p. 080403 (2001).

26. J. M. McNamara, T. Jeltes, A. S. Tychkov, W. Hogervorst and W. Vassen, *Phys. Rev. Lett.* **97**, p. 080404 (2006).
27. T. Fukuhara, Y. Takasu, M. Kumakura and Y. Takahashi, *Phys. Rev. Lett.* **98**, p. 030401 (2007).
28. S. Taie, Y. Takasu, S. Sugawa, R. Yamazaki, T. Tsujimoto, R. Murakami and Y. Takahashi, *Phys. Rev. Lett.* **105**, p. 190401 (2010).
29. B. J. DeSalvo, M. Yan, P. G. Mickelson, Y. N. Martinez de Escobar and T. C. Killian, *Phys. Rev. Lett.* **105**, p. 030402 (2010).
30. M. K. Tey, S. Stellmer, R. Grimm and F. Schreck, *Phys. Rev. A* **82**, p. 011608 (2010).
31. C. Chin, R. Grimm, P. Julienne and E. Tiesinga, *Rev. Mod. Phys.* **82**, 1225 (2010).
32. C. A. Regal, M. Greiner and D. S. Jin, *Phys. Rev. Lett.* **92**, p. 040403 (2004).
33. D. Jaksch, C. Bruder, J. I. Cirac, C. W. Gardiner and P. Zoller, *Phys. Rev. Lett.* **81**, 3108 (1998).
34. M. Greiner, O. Mandel, T. Esslinger, T. Hänsch and I. Bloch, *Nature* **415**, 39 (2002).
35. R. Jördens, N. Strohmaier, K. Günter, H. Moritz and T. Esslinger, *Nature* **455**, 204 (2008).
36. U. Schneider, L. Hackermueller, S. Will, T. Best, I. Bloch, T. A. Costi, R. W. Helmes, D. Rasch and A. Rosch, *Science* **322**, 1520 (2008).
37. M. Köhl, H. Moritz, T. Stöferle, K. Günter and T. Esslinger, *Phys. Rev. Lett.* **94**, p. 080403 (2005).
38. R. Feynman, *International journal of theoretical physics* **21**, 467 (1982).
39. A. Rapp, G. Zaránd, C. Honerkamp and W. Hofstetter, *Phys. Rev. Lett.* **98**, p. 160405 (2007).
40. M. A. Cazalilla, A. F. Ho and M. Ueda, *New J. Phys.* **11**, p. 103033 (2009).
41. J. Struck, C. Olschlager, R. Le Targat, P. Soltan-Panahi, A. Eckardt, M. Lewenstein, P. Windpassinger and K. Sengstock, *Science* **333**, 996 (2011).
42. Y.-J. Lin, K. Jiménez-García and I. B. Spielman, *Nature* **471**, 83 (2011).
43. M. Aidelsburger, M. Atala, S. Nascimbène, S. Trotzky, Y.-A. Chen and I. Bloch, *Phys. Rev. Lett.* **107**, p. 255301 (2011).
44. M. Kitagawa, K. Enomoto, K. Kasa, Y. Takahashi, R. Ciuryło, P. Naidon and P. S. Julienne, *Phys. Rev. A* **77**, p. 012719 (2008).
45. K. Shibata, S. Kato, A. Yamaguchi, S. Uetake and Y. Takahashi, *Applied Physics B: Lasers and Optics* **97**, 753 (2009).
46. A. J. Daley, M. M. Boyd, J. Ye and P. Zoller, *Phys. Rev. Lett.* **101**, p. 170504 (2008).
47. A. Yamaguchi, S. Uetake, S. Kato, H. Ito and Y. Takahashi, *New Journal of Physics* **12**, p. 103001 (2010).
48. P. O. Fedichev, Y. Kagan, G. V. Shlyapnikov and J. T. M. Walraven, *Phys. Rev. Lett.* **77**, 2913 (1996).
49. R. Ciuryło, E. Tiesinga and P. S. Julienne, *Phys. Rev. A* **71**, p. 030701 (2005).
50. K. Enomoto, K. Kasa, M. Kitagawa and Y. Takahashi, *Phys. Rev. Lett.* **101**, p. 203201 (2008).

51. R. Yamazaki, S. Taie, S. Sugawa and Y. Takahashi, *Phys. Rev. Lett.* **105**, p. 050405 (2010).
52. F. Gerbier, A. Widera, S. Fölling, O. Mandel, T. Gericke and I. Bloch, *Phys. Rev. Lett.* **95**, p. 050404 (2005).
53. S. Sugawa, K. Inaba, S. Taie, R. Yamazaki, M. Yamashita and Y. Takahashi, *Nature Physics* **7**, 642 (2011).
54. K. M. Jones, E. Tiesinga, P. D. Lett and P. S. Julienne, *Rev. Mod. Phys.* **78**, 483 (2006).
55. T. Stöferle, H. Moritz, C. Schori, M. Köhl and T. Esslinger, *Phys. Rev. Lett.* **92**, p. 130403 (2004).
56. D. Greif, L. Tarruell, T. Uehlinger, R. Jördens and T. Esslinger, *Phys. Rev. Lett.* **106**, p. 145302 (2011).
57. R. C. Richardson, *Rev. Mod. Phys.* **69**, 683 (1997).
58. M. Cramer, *Phys. Rev. Lett.* **106**, p. 215302 (2011).
59. L. Hackermuller, U. Schneider, M. Moreno-Cardoner, T. Kitagawa, T. Best, S. Will, E. Demler, E. Altman, I. Bloch and B. Paredes, *Science* **327**, 1621 (2010).
60. I. Titvinidze, M. Snoek and W. Hofstetter, *Phys. Rev. Lett.* **100**, p. 100401 (2008).
61. M. Lewenstein, L. Santos, M. A. Baranov and H. Fehrmann, *Phys. Rev. Lett.* **92**, p. 050401 (2004).
62. A. Eckardt and M. Lewenstein, *Phys. Rev. A* **82**, p. 011606 (2010).
63. J.-S. Bernier, C. Kollath, A. Georges, L. De Leo, F. Gerbier, C. Salomon and M. Köhl, *Phys. Rev. A* **79**, p. 061601 (2009).
64. M. Lubasch, V. Murg, U. Schneider, J. I. Cirac and M.-C. Bañuls, *Phys. Rev. Lett.* **107**, p. 165301 (2011).
65. J. Catani, G. Barontini, G. Lamporesi, F. Rabatti, G. Thalhammer, F. Minardi, S. Stringari and M. Inguscio, *Phys. Rev. Lett.* **103**, p. 140401 (2009).
66. K. R. A. Hazzard, V. Gurarie, M. Hermele and A. M. Rey, p. Preprint at http://arxiv.org/abs/1011.0032 (2010).

UNIVERSALITY OF INTEGRABLE MODEL: BAXTER'S T-Q EQUATION, $SU(N)/SU(2)^{N-3}$ CORRESPONDENCE AND Ω-DEFORMED SEIBERG-WITTEN PREPOTENTIAL

TA-SHENG TAI*

*Interdisciplinary Graduate School of Science and Engineering,
Kinki University, 3-4-1 Kowakae, Higashi-Osaka, Osaka 577-8502, Japan*
*[b] Osaka City University Advanced Mathematical Institute,
3-3-138 Sugimoto, Sumiyoshi-ku, Osaka 558-8585, Japan*

Integrable models in two dimensions are well-studied. Their appearance proved to be so universal in various kinds of topics including 2D conformal field theory, 3D Chern-Simons theory, to name a few. We present how 4D supersymmetric gauge theory also gets related to spin-chain models.

More precisely, we study Baxter's T-Q equation of XXX spin-chain models under the semiclassical limit where an intriguing $SU(N)/SU(2)^{N-3}$ correspondence emerges. That is, two kinds of 4D $\mathcal{N}=2$ superconformal field theories having the above different gauge groups are encoded simultaneously in one Baxter's T-Q equation which captures their spectral curves. For example, while one is $SU(N_c)$ with $N_f = 2N_c$ flavors the other turns out to be $SU(2)^{N_c-3}$ with N_c hyper-multiplets ($N_c > 3$). It is seen that the corresponding Seiberg-Witten differential supports our proposal.

1. Introduction and summary

Recently, there have been new insights into the duality between integrable systems and 4D $\mathcal{N}=2$ gauge theories. In 1–3 Nekrasov and Shatashvili (NS) have found that Yang-Yang functions as well as Bethe Ansatz equations of a family of integrable models are indeed encoded in a variety of Nekrasov's partition functions[4,5] restricted to the two-dimensional Ω-background[a]. As a matter of fact, this mysterious correspondence can further be extended to the full Ω-deformation in view of the birth of AGT conjecture.[11] Let us briefly refine the latter point.

---

*E-mail address: tasheng@alice.math.kindai.ac.jp
[a]See also recent 6–10 which investigated XXX spin-chain models along this line.

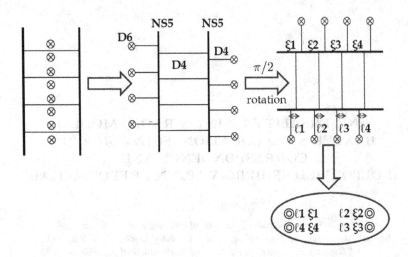

Fig. 1. Main idea: $SU(N)/SU(2)^{N-3}$ correspondence LHS: M-theory curve of $SU(4)$ $N_f = 8$ Yang-Mills theory embedded in $\mathbf{C} \times \mathbf{C}^*$ parameterized by (u,w) ($w = \exp(-s/R)$, $R = \ell_s g_s$: M-circle radius); RHS: spin-chain variables (ξ_n, ℓ_n) labeling (coordinate, weight) of each puncture on $\mathbf{CP}^1$ (but indicating each flavor D6-brane location along u-plane of LHS)

AGT claimed that correlators of primary states in Liouville field theory (LFT) can get re-expressed in terms of Nekrasov's partition function Z_{Nek} of 4D $\mathcal{N} = 2$ quiver-type $SU(2)$ superconformal field theories (SCFTs). In particular, every Riemann surface $C_{g,n}$ (whose doubly-sheeted cover is called Gaiotto curve[12]) on which LFT dwells is responsible for one specific SCFT called $\mathcal{T}_{g,n}(A_1)$ such that the following equality

$$\text{conformal block w.r.t. } C_{g,n} = \text{instanton part of } Z_{\text{Nek}}\Big(\mathcal{T}_{g,n}(A_1)\Big)$$

holds. Because of $\epsilon_1 : \epsilon_2 = b^{-1} : b^{11}$ the one-parameter version ($\epsilon_2 = 0$) of AGT conjecture directly leads to the semiclassical LFT as $b \to 0$. Quote further the geometric Langlands correspondence[13] which associates Gaudin integrable models on the projective line with LFT at $b \to 0$. It is then plausible to put both proposals of NS and AGT into one unified scheme.

In this letter, we add a new element into the above 2D/4D correspondence. Starting from Baxter's T-Q equation of XXX spin-chain models we found a novel interpretation of it. That is, under the semiclassical limit it possesses two aspects simultaneously. It describes
- 4D $\mathcal{N} = 2$ $SU(N_c)$ Yang-Mills with $N_f = 2N_c$ flavors, $\mathcal{T}_{0,4}(A_{N_c-1})$, on the one hand and

- $SU(2)^{N_c-3}$ ($N_c > 3$) quiver-type Yang-Mills with N_c (four fundamental and $N_c - 4$ bi-fundamental) hyper-multiplets, $\mathcal{T}_{0,N_c}(A_1)$, on the other hand. It is helpful to have a rough idea through Fig. 1. Pictorially, $C_{0,4}$ for $\mathcal{T}_{0,4}(A_1)$ in RHS results from the encircled part in LHS after a $\pi/2$-rotation.

In other words, the conventional Type IIA Seiberg-Witten (SW) curve in fact contains another important piece of information while seen from (u,v)-space $(u = x^4 + ix^5, v = x^7 + ix^8)$[b]. Here, "$\pi/2$-rotation" just means that SW differentials of two theories thus yielded are connected by exchanging $(u, s = x^6 + ix^{10})$.

This quite unexpected phenomenon will be explained later by combining a couple of topics, say, Bethe Ansatz, Gaudin model and Liouville theory. Roughly speaking, the spin-chain variable ℓ (ξ), highest weight (shifting parameter), is responsible for m (q) of RHS in Fig. 1. As summarized in Table 1, N_c Coulomb moduli $\xi \in \mathbf{C}$ (one overall $U(1)$ factor) are mapped to $N_c - 3$ gauge coupling constants $q = \exp(2\pi i \tau) \in \mathbf{C}^*$ where three of them are fixed to $(0, 1, \infty)$ on $\mathbf{C}^*$. Those entries marked by $\odot$ do not have direct comparable counterparts.

# of UV parameter	LHS	RHS
Coulomb moduli	$N_c - 1$ (ξ)	$\odot N_c - 3$ (a)
bare flavor mass	N_c ($\xi \pm \ell$)	N_c (m)
gauge coupling	$\odot 1 \exp(\dfrac{\Delta x^6 + i\Delta x^{10}}{R})$	$N_c - 3$ (q)

We organize this letter as follows. Sec. 2 is devoted to a further study of Fig. 2 on which our main idea Fig. 1 is based. Then Sec. 3 unifies three elements: Gaudin model, LFT and matrix model as shown in Fig. 3. Finally, in Sec. 4 we complete our proposal by examining λ_{SW} (SW differential) and shortly discuss XYZ Gaudin models.

2. XXX spin chain

Baxter's T-Q equation[15,16] plays an underlying role in various spin-chain models. On the other hand, it has long been known that the low-energy Coulomb sectors of $\mathcal{N} = 2$ gauge theories are intimately related to a variety of integrable systems.[17-23] Here, by integrable model (or solvable model)

---

[b]This aspect of $\mathcal{N} = 2$ curves is also stressed in 14.

we mean that there exists some spectral curve which gives enough integrals of motion (or conserved charges). In the case of $\mathcal{N}=2$ $SU(N_c)$ Yang-Mills theory with N_f fundamental hyper-multiplets, its SW curve[24,25] is identified with the spectral curve of an inhomogeneous periodic Heisenberg XXX spin chain on N_c sites:

$$w + \frac{1}{w} = \frac{P_{N_c}(u)}{\sqrt{Q_{N_f}(u)}}. \tag{1}$$

Here, two polynomials P_{N_c} and Q_{N_f} encode respectively parameters of $\mathcal{N}=2$ vector- and hyper-multiplets. Meanwhile, the meromorphic SW differential $\lambda_{SW} = u d \log w$ provides a set of "special coordinates" through its period integrals (see Table 1):

$$\xi_n = \oint_{\alpha_n} \lambda_{SW}, \qquad \frac{\partial \mathcal{F}_{SW}}{\partial \xi_n} = \xi_n^D = \oint_{\beta_n} \lambda_{SW}, \qquad \xi_n \pm \ell_n = \oint_{\gamma_n^\pm} \lambda_{SW} \tag{2}$$

where $\mathcal{F}_{SW}$ is the physical prepotential.

2.1. Baxter's T-Q equation

Indeed, (1) arises from (up to $w \to \sqrt{Q_{N_f}} w$)

$$\det(w - T(u)) = 0 \quad \to \quad w^2 - \operatorname{tr} T(u) w + \det T(u) = 0,$$

$T(u)$: 2×2 monodromy matrix,

$$\det T(u) = Q_{N_f}(u) = \prod_{n=1}^{N_c}(u - m_n^-)(u - m_n^+), \qquad m_n^\pm = \xi_n \pm \ell_n.$$

$\operatorname{tr} T(u) = t(u) = P_{N_c}(u) = \langle \det(u - \Phi) \rangle$, transfer matrix, encodes the quantum vev of the adjoint scalar field Φ. In fact, (1) belongs to the *conformal* case where $N_f = 2N_c$ bare flavor masses are indicated by $m_n^\pm$. It is time to quote Baxter's T-Q equation:

$$t(u) Q(u) = \Delta_+(u) Q(u - 2\eta) + \Delta_-(u) Q(u + 2\eta). \tag{3}$$

Some comments follow:
- η is Planck-like and ultimately gets identified with ϵ_1 (one of two Ω-background parameters) in Sec. 4.
- As a matter of fact, (3) boils down to (1) (up to $w \to \sqrt{Q_{N_f}} w$) as $\eta \to 0$. Curiously, then its λ_{SW} signals the existence of another advertised $\mathcal{N}=2$ theory. The situation is pictorially shown in Fig. 1.
- Remark again that SW differentials of two theories are connected by exchanging two holomorphic coordinates (u, s) but their M-lifted[26] Type

IIA brane configurations[c] are not. Instead, the $\pi/2$-rotated part is closely related to $\mathcal{N} = 2$ Gaiotto's curve. A family of quiver-type $SU(2)$ SCFTs $\mathcal{T}_{0,n}(A_1)$ discovered by Gaiotto[12] is hence made contact with.

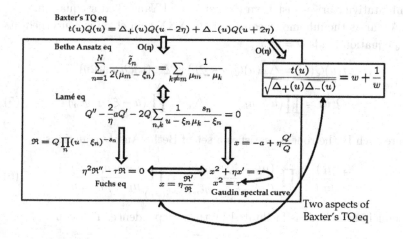

Fig. 2. Mathematical description of Fig. 1 Up to $\mathcal{O}(\eta)$, Baxter's T-Q equation and Bethe Ansatz equations of it describe two kinds of $\mathcal{N} = 2$ gauge theories which however are related by one λ_{SW}

2.2. More detail

Let us refine the above argument. Consider a quantum spin-chain built over an N-fold tensor product $\mathcal{H} = \otimes_{n=1}^{N} V_n$. In other words, at each site labeled by n we assign an irreducible representation V_n of $\mathfrak{sl}_2$ which is $(\ell_n + 1)$-dimensional where $\ell_n = 0, 1, 2, \cdots$. Therefore, ℓ_n denotes the highest weight. Within the context of QISM[d], monodromy and transfer matrices are defined respectively by

$$T(u) = \begin{pmatrix} A_N(u) & B_N(u) \\ C_N(u) & D_N(u) \end{pmatrix} = L_N(u - \xi_N) \cdots L_1(u - \xi_1), \quad (4)$$

$$\hat{t}(u) = A_N(u) + D_N(u).$$

---

[c]In 27 this symmetry has been notified in the context of Toda-chain models because two kinds of Lax matrices exist there.
[d]Quantum Inverse Scattering Method (QISM) was formulated in 1979-1982 in St. Petersburg Steklov Mathematical Institute by Faddeev and many of his students. We thank Petr Kulish for informing us of this fact.

V_n is acted on by the n-th Lax operator L_n. By *inhomogeneous* one means that the spectral parameter u has been shifted by ξ. Conventionally, $\hat{t}(u)$ or its eigenvalue $t(u)$ is called generating function because a series of conserved charges can be extracted from its coefficients owing to $[\hat{t}(u), \hat{t}(v)] = 0$. The commutativity arises just from the celebrated Yang-Baxter equation.

As far as the inhomogeneous periodic XXX spin chain is concerned, its T-Q equation reads ($\ell = \eta\tilde{\ell}$)

$$t(u)Q(u) = \triangle_+(u)Q(u - 2\eta) + \triangle_-(u)Q(u + 2\eta),$$

$$Q(u) = \prod_{k=1}^{K}(u - \mu_k), \qquad \triangle_\pm = \prod_{n=1}^{N}(u - \xi_n \pm \eta\tilde{\ell}_n) \qquad (5)$$

where each Bethe root μ_k satisfies a set of Bethe Ansatz equations:

$$\frac{\triangle_+(\mu_k)}{\triangle_-(\mu_k)} = \prod_{n=1}^{N}\frac{(\mu_k - \xi_n + \eta\tilde{\ell}_n)}{(\mu_k - \xi_n - \eta\tilde{\ell}_n)} = \prod_{l(\neq k)}^{K}\frac{\mu_k - \mu_l + 2\eta}{\mu_k - \mu_l - 2\eta}. \qquad (6)$$

A semiclassical limit is facilitated by the η dependence. Through

$$\frac{t(u)}{\sqrt{\triangle_+ \triangle_-}} = \frac{Q(u - 2\eta)}{Q(u)}\sqrt{\frac{\triangle_+}{\triangle_-}} + \frac{Q(u + 2\eta)}{Q(u)}\sqrt{\frac{\triangle_-}{\triangle_+}} \qquad (7)$$

and omitting $\mathcal{O}(\eta^2)$, we have

$$\frac{t(u)}{\sqrt{\triangle_+(u)\triangle_-(u)}} = w + \frac{1}{w}, \qquad w \equiv \sqrt{\frac{\triangle_+}{\triangle_-}}(1 - 2\eta\frac{Q'}{Q}) \qquad (8)$$

which exactly reduce to (1). Throughout this letter, $(\eta, \tilde{\ell})$ while kept finite are, respectively, small and large.

From now on, we call $\lambda_{SW} = ud\log w \equiv \lambda^\eta_{SW}$ "η-deformed" SW differential as in 28,29:

$$\lambda^\eta_{SW} = 2\eta ud\left(\frac{\Psi'}{\Psi}\right) + \mathcal{O}(\eta^2), \qquad \Psi = \frac{1}{Q(u)}\prod_n (u - \xi_n)^{\tilde{\ell}_n/2}. \qquad (9)$$

Also, up to $\mathcal{O}(\eta)$ (6) reads

$$\sum_{n=1}^{N}\frac{\tilde{\ell}_n}{2(\mu_k - \xi_n)} = \sum_{l(\neq k)}^{K}\frac{1}{(\mu_k - \mu_l)}. \qquad (10)$$

That λ^η_{SW} looks strikingly similar to (10) signals the existence of RHS in Fig. 1. Fig. 2 outlines our logic. One will find that λ^η_{SW} naturally emerges as the holomorphic one-form of Gaudin's spectral curve which captures

Gaiotto's curve for $\mathcal{T}_{0,N}(A_1)$. In what follows, our goal is to show that λ_{SW}^{η} does reproduce the ϵ_1-deformed SW prepotential w.r.t. $\mathcal{T}_{0,N}(A_1)$.

Several comments follow:

• In M-theory D6-branes correspond to singular loci of $XY = \triangle_+(u)\triangle_-(u)$. This simply means that one incorporates flavors via replacing a flat $\mathbf{R}^4$ over (u,s) by a resolved $\mathbf{A}_{2N_c-1}$-type singularity.

• Without flavors (i.e. turning off ℓ) $\int \lambda_{SW}^{\eta}$ looks like a logarithm of the usual Vandermonde. This happens in the familiar Dijkgraaf-Vafa story[30–32] without any tree-level potential which brings $\mathcal{N} = 2$ pure Yang-Mills to $\mathcal{N} = 1$ descendants.

• Surely, this intuition is noteworthy in view of (10) which manifests itself as the saddle-point condition within the context of matrix models. To pursue this interpretation, one should regard μ's as diagonal elements of $\mathcal{M}$ (Hermitian matrix of size $K \times K$). Besides, the tree-level potential now obeys

$$\mathcal{W}'(x) = -\sum_{n=1}^{N} \frac{\ell_n}{(x-\xi_n)}.$$

In other words, we are equivalently dealing with "$\mathcal{N} = 2$" Penner-type matrix models which have been heavily investigated recently in connection with AGT conjecture due to [33]. We will return to these points soon.

3. XXX Gaudin model

Momentarily, we focus on another well-studied integrable model: XXX Gaudin model. The essential difference between Heisenberg and Gaudin models amounts to the definition of their generating functions. Following Fig. 3 we want to explain two important aspects of Gaudin's spectral curve.

3.1. *RHS of Fig. 3*

Expanding around small η, we yield

$$L_n(u) = 1 + 2\eta \mathcal{L}_n + O(\eta^2), \tag{11}$$

$$T(u) = 1 + 2\eta \mathcal{T} + \eta^2 \mathcal{T}^{(2)} + O(\eta^3), \tag{12}$$

$$t(u) = 1 + \eta^2 \mathrm{tr}\mathcal{T}^{(2)} + O(\eta^3), \tag{13}$$

$$\tau(u) \equiv \frac{1}{2}\eta^2 \mathrm{tr}\mathcal{T}^2, \qquad \mathcal{T} = \sum_n \mathcal{L}_n = \begin{pmatrix} A(u) & B(u) \\ C(u) & -A(u) \end{pmatrix} \tag{14}$$

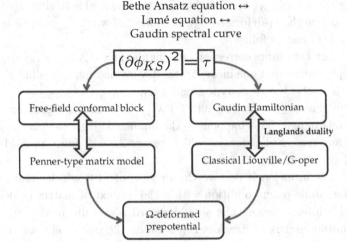

Fig. 3. Flow chart of Sec. 3

where
$$A(u) = \sum_{n=1}^{N} \frac{J_n^z}{u-\xi_n}, \quad B(u) = \sum_{n=1}^{N} \frac{J_n^-}{u-\xi_n}, \quad C(u) = \sum_{n=1}^{N} \frac{J_n^+}{u-\xi_n} \quad (15)$$

while $\vec{J} = (J^z, J^\pm)$ represents generators of $\mathfrak{sl}_2$ Lie algebra. Instead of $\operatorname{tr} \mathcal{T}^{(2)}$ ($\operatorname{tr} \mathcal{T} = 0$) the generating function adopted is ($s = \widetilde{\ell}/2 = \ell/2\eta$)

$$\tau(u) = \sum_{n=1}^{N} \left\{ \frac{\eta^2 s_n(s_n+1)}{(u-\xi_n)^2} + \frac{c_n}{u-\xi_n} \right\},$$

$$c_n = \sum_{i \neq n}^{N} \frac{2\eta^2 \vec{J}_n \cdot \vec{J}_i}{\xi_n - \xi_i}, \qquad \vec{J}_n \cdot \vec{J}_n = s_n(s_n+1). \quad (16)$$

Conventionally, c_n's are called Gaudin Hamiltonians which commute with one another as a result of the classical Yang-Baxter equation.

$$\Sigma: \quad x^2 = \tau(u) \subset T^*C$$

is the N-site Gaudin spectral curve, a doubly-sheeted cover of $C \equiv \mathbf{CP}^1 \backslash \{\xi_1, \cdots, \xi_N\}$.

According to the geometric Langlands correspondence[e], c_n's give exactly

---

[e]See also 34–36.

accessory parameters of a G-oper:

$$\mathcal{D} = -\partial_z^2 + \sum_{n=1}^{N} \frac{\delta_n}{(z-\xi_n)^2} + \sum_{n=1}^{N} \frac{\widetilde{c}_n}{z-\xi_n}, \qquad \delta = s(s+1), \qquad c = \eta^2 \widetilde{c}$$

defined over $C = \mathbf{CP}^1 \setminus \{\xi_1, \cdots, \xi_N\}$. The non-singular behavior of $\mathcal{D}$ is ensured by imposing

$$\sum_{n=1}^{N} \widetilde{c}_n = 0, \qquad \sum_{n=1}^{N} (\xi_n \widetilde{c}_n + \delta_n) = 0, \qquad \sum_{n=1}^{N} (\xi_n^2 \widetilde{c}_n + 2\xi_n \delta_n) = 0.$$

Certainly, one soon realizes that $\tau(u)$ here is nothing but the holomorphic LFT (2,0) stress-tensor as the central charge $1 + 6Q^2$ goes to infinity (or $b \to 0$). Namely,

$$\eta^{-2}\tau \equiv \frac{1}{2}\partial_z^2 \varphi_{cl} - \frac{1}{4}(\partial_z \varphi_{cl})^2 = \sum_{n=1}^{N} \frac{\delta_n}{(z-\xi_n)^2} + \sum_{n=1}^{N} \frac{\widetilde{c}_n}{z-\xi_n}.$$

In terms of LFT, the second equality comes from Ward identity of the stress-tensor $T_L = \frac{1}{2}Q\partial_z^2\phi - \frac{1}{4}(\partial_z\phi)^2$ inserted in $\langle \prod_n V_{\alpha_n} \rangle$ subject to $b \to 0$. Here, $V_\alpha = \exp(2\alpha\phi)$ denotes the primary field ($\Delta_\alpha = \alpha(Q-\alpha)$, $Q = b + b^{-1}$). As $b \to 0$,

$$\langle (-T_L) \prod_n V_{\alpha_n} \rangle = \int \mathcal{D}\phi \exp(-\mathbf{S}_{\text{tot}})(-T_L) \to \exp(-\frac{1}{b^2}\mathcal{S}_{\text{tot}}[\varphi_{cl}])\frac{1}{\eta^2 b^2}\tau$$

such that for the unique saddle-point to $\mathcal{S}_{\text{tot}}[\varphi]$ one has (Polyakov conjecture)

$$\widetilde{c}_n = \frac{\partial \mathcal{S}_{\text{tot}}[\varphi_{cl}]}{\partial \xi_n}, \qquad |\widetilde{\alpha}_n| = b|\alpha_n| = s_n \qquad (17)$$

where on a large disk Γ

$$\mathbf{S}_{\text{tot}} = \int_\Gamma d^2 z \Big(\frac{1}{4\pi}|\partial_z\phi|^2 + \mu e^{2b\phi}\Big) + \text{boundary terms}, \qquad \mathbf{S}_{\text{tot}}[\phi] = \frac{1}{b^2}\mathcal{S}_{\text{tot}}[\varphi].$$

Note that φ_{cl} satisfies Liouville's equation and is important during uniformizing Riemann surfaces with constant negative curvature. Usually, $\widetilde{\alpha} = b\alpha$ is kept fixed during $b \to 0$. It is necessary that $\eta = \hbar/b$ due to $b|\alpha| = \ell/2\eta$. This confirms in advance $\eta = \epsilon_1$ due to AGT dictionary.

3.2. LHS of Fig. 3

As shown in 37, $\tau(u)$ has another form in terms of the eigenvalue $a(u)$ of $A(u)$[f]:

$$\tau(u) = a^2 - \eta a' - 2\eta \sum_k \frac{a(u) - a(\mu_k)}{u - \mu_k}, \qquad a(u) \equiv \sum_{n=1}^N \frac{\eta s_n}{(u - \xi_n)} \qquad (18)$$

with μ_k's being Bethe roots. This expression is extremely illuminating in connection with Penner-type matrix models. Borrowing $Q(u)$ from (5) and defining

$$\Re(u) \equiv Q(u) \exp\left(-\frac{1}{\eta}\int^u a(y)dy\right) = \prod_k (u - \mu_k) \prod_n (u - \xi_n)^{-s_n}, \quad (19)$$

we can verify that there holds

$$\eta x' + x^2 = \tau, \qquad x(u) = \eta \frac{\Re'(u)}{\Re(u)} = -a + \sum_k \frac{\eta}{u - \mu_k}. \qquad (20)$$

This is the so-called Lamé equation in disguise. Equivalently, $\Re(u)$ solves a Fuchs-type equation $(\eta^2 \partial_u^2 - \tau(u))\Re(u) = 0$ with N regular singularities on $\mathbf{CP}^1$.

Compared with x^2, $\eta x'$ becomes subleading. Further getting rid of $\eta x'$, we arrive at Gaudin's spectral curve

$$x^2 = \tau. \qquad (21)$$

In view of (20), it is tempting to introduce ϕ_{KS}, i.e. Kodaira-Spencer field w.r.t. Z^M defined in (24). That is,

$$2x \equiv \partial \phi_{KS} = \mathcal{W}' + 2\eta \operatorname{tr}\left\langle \frac{1}{u - \mathcal{M}} \right\rangle_{Z^M}. \qquad (22)$$

Subsequently, (21) becomes precisely the spectral curve of Z^M. Remark that

$$\oint \partial \phi_{KS} du = -\oint \lambda_{SW}^\eta \qquad (23)$$

up to a total derivative term. Additionally, it is well-known that from the period integral (23) one yields the tree-level free energy $\mathcal{F}_0$ of Z^M:

$$Z^M = \int_{K \times K} \mathcal{DM} \exp\left[\frac{1}{\eta}\mathcal{W}(\mathcal{M})\right], \quad \mathcal{W}' = -\sum_{n=1}^N \frac{\ell_n}{(u - \xi_n)}, \quad K\eta = \text{fixed}. \qquad (24)$$

---

[f]We hope that readers will not confuse $a(u)$ here with a denoting Coulomb moduli.

Of course, the saddle-point of Z^M is dictated by (10). We want to display in Sec. 4 that $\mathcal{F}_0$ is surely related to $\mathcal{T}_{0,N}(A_1)$. In view of (23), we refer to this as the advertised $SU(N)/SU(2)^{N-3}$ ($N > 3$) correspondence.

For Gaudin's spectral curve, due to $(x, u) \in \mathbf{C} \times \mathbf{C}^*$ we introduce $v = xu$ such that xdu here and the former $ud\log w$ of $\mathcal{T}_{0,4}(A_{N-1})$ look more symmetrical. Moreover, the $\pi/2$-rotation noted in Fig. 1 is only pictorial otherwise one naively has $SU(N)/SU(2)^{N-1}$ correspondence instead[g].

3.2.1. Free-field representation

As another crucial step, we rewrite Z^M in terms of a multi-integral over diagonal elements of $\mathcal{M}$:

$$Z^M \equiv \oint dz_1 \cdots \oint dz_K \prod_{i<j}(z_i - z_j)^2 \prod_{i,n}(z_i - \xi_n)^{-\widetilde{\ell}_n} \prod_{n<m}(\xi_n - \xi_m)^{\widetilde{\ell}_n \widetilde{\ell}_m/2}.$$
(25)

A constant term involving only ξ's is multiplied by hand. This form then realizes a chiral conformal block of N LFT primary fields via Feigin-Fuchs free-field representation. Notably, the charge balance condition is respected in the presence of background charge Q via inserting K screening operators $\oint dz \exp 2b^{-1}\phi(z)$ of zero conformal weight. Also, the free propagator $\langle \phi(z_1)\phi(z_2)\rangle_{\text{free}} = -\log(z_1 - z_2)^{1/2}$ is used.

Assume the genus expansion $Z^M = \exp(\eta^{-2}\mathcal{F}_0 + \cdots)$ and

$$\lim_{b\to 0} \log \left\langle V_{-\widetilde{\ell}_1/2b}(\xi_1) \cdots V_{-\widetilde{\ell}_N/2b}(\xi_N) \right\rangle_{\text{conformal block}} = -b^{-2}\widetilde{F}.$$

$\widetilde{F}$ named *classical* conformal block appeared in the pioneering work of Zamolodchikov and Zamolodchikov.[38] Based on the above discussion, one can anticipate that $\eta^2 \widetilde{F} = \mathcal{F}_0$. Next, to identify $\mathcal{F}_0$ with the Ω-deformed SW prepotential for $\mathcal{T}_{0,N}(A_1)$ serves as the last step towards completing our proposal.

4. Application and discussion

Without loss of generality, we examine a concrete example: $N = 4$. As a result, $\mathcal{F}_0$ generated by the period integral of $\lambda_{SW}^\eta = -\partial \phi_{KS} du$ is indeed the very ϵ_1-deformed SW prepotential of $\mathcal{T}_{0,4}(A_1)$. Quote the known $\tau(u)$

---

[g]We thank Yuji Tachikawa for his comment on this point.

for $N = 4$ from 38:

$$\eta^{-2}\tau(u) = \frac{\delta_1}{u^2} + \frac{\delta_2}{(u-q)^2} + \frac{\delta_3}{(1-u)^2} + \frac{\delta_1 + \delta_2 + \delta_3 - \delta_4}{u(1-u)} + \frac{q(1-q)\tilde{c}(q)}{u(u-q)(1-u)}. \tag{26}$$

Via projective invariance q represents the cross-ratio of four marked points $(\xi_1, \xi_2, \xi_3, \xi_4) \equiv (0, 1, q, \infty)$ on $\mathbf{CP}^1$.

The residue of τ around $u = 1$ is ($v = xu$)

$$q\tilde{c}(q) = -\eta^{-2} \oint ux^2 du = -\frac{1}{2}\eta^{-2} \oint v\lambda_{SW}^\eta = q\frac{\partial}{\partial q}\widetilde{F}_{\delta,\delta_n}(q), \quad (n = 1, \cdots, 4), \tag{27}$$

$$\widetilde{F}_{\delta,\delta_n}(q) = (\delta - \delta_1 - \delta_2)\log q + \frac{(\delta + \delta_1 - \delta_2)(\delta + \delta_3 - \delta_4)}{2\delta}q + \mathcal{O}(q^2)$$

where Polyakov's conjecture (17) is applied in the last equality of (27). Notice that only the holomorphic $\widetilde{F}$ in $\mathcal{S}_{\text{tot}}$ survives $\partial/\partial q$. Conversely, by taking into account the stress-tensor nature of the spectral curve $(\partial \phi_{KS})^2 = 4\tau$ in Hermitian matrix models, $\widetilde{F}$ can be replaced by $\mathcal{F}_0$ as a result of Virasoro algebra. This observation supports the above $\eta^2 \widetilde{F} = \mathcal{F}_0$.

Finally, we need another ingredient: Matone's relation.[39–41] As is proposed in 28,29,42, the ϵ_1-deformed version is

$$\langle \operatorname{tr} \Phi^2 \rangle_{\epsilon_1} = 2\bar{q}\partial_{\bar{q}}W, \qquad \bar{q} = \exp(2\pi i \bar{\tau}_{UV}) \tag{28}$$

for, say, $\mathcal{N} = 2$ $\mathcal{T}_{0,4}(A_1)$ theory where

$$\frac{1}{\epsilon_1 \epsilon_2} W(\epsilon_1) \equiv \lim_{\epsilon_2 \to 0} \log Z_{\text{Nek}}(a, \vec{m}, \bar{q}, \epsilon_1, \epsilon_2),$$

a : UV vev of Φ, $\qquad \vec{m}$: four bare flavor masses.

Now, (27) and (28) together manifest λ_{SW}^η as the ϵ_1-deformed SW differential for $\mathcal{T}_{0,4}(A_1)$ if there holds

$$\frac{1}{b^2}\widetilde{F}_{\delta,\delta_n}(q) = \frac{1}{\hbar^2}\mathcal{F}_0 = \lim_{\epsilon_2 \to 0} \frac{1}{\epsilon_1 \epsilon_2} W(\epsilon_1) \tag{29}$$

under $q = \bar{q}$, $\epsilon_1 = \eta$ and $\epsilon_1 \epsilon_2 = \hbar^2$. In fact, (29) has already been verified in 43. To conclude, by examining λ_{SW}^η we have found that Baxter's T-Q equation encodes simultaneously two kinds of $\mathcal{N} = 2$ theories, $\mathcal{T}_{0,N}(A_1)$ and $\mathcal{T}_{0,4}(A_{N-1})$. We call this remarkable property $SU(N)/SU(2)^{N-3}$ correspondence.

Theory of RHS in Fig. 1 ($N = 4$)
$q = \dfrac{(\xi_1 - \xi_3)(\xi_2 - \xi_4)}{(\xi_2 - \xi_3)(\xi_1 - \xi_4)}$
$\epsilon_1 m \leftrightarrow \ell$
$(\epsilon_1, \epsilon_2) = (\eta, 0)$
$a = \oint_\alpha \lambda_{SW}^\eta$
$\epsilon_1^2 \dfrac{\partial \widetilde{F}}{\partial a} = \oint_\beta \lambda_{SW}^\eta$

4.1. Discussion

• Based on (8) and (9), we have at the level of λ_{SW}^η

$$\log w = 2\eta \frac{\Psi'}{\Psi}, \qquad w = \frac{A + \sqrt{A^2 - 4}}{2}, \qquad A = \frac{P_{N_c}}{\sqrt{Q_{N_f}}}. \qquad (30)$$

Namely, all quantum $SU(N_c)$ Coulomb moduli encoded inside $P_{N_c}(u) \equiv \langle \det(u - \Phi) \rangle$ are determined by using spin-chain variables (η, ξ, ℓ). This fact is consistent with (2).

• Besides, from Table 2 we find that the transformation between $\mathcal{F}_{SW}$ in (2) and $\widetilde{F}$ is quite complicated. Although sharing the same SW differential (up to a total derivative term), two theories have diverse IR dynamics because both of their gauge group and matter content differ. To pursue a concrete interpolation between them is under investigation.

4.2. XYZ Gaudin model

There are still two other Gaudin models, say, trigonometric and elliptic ones. Let us briefly discuss the elliptic type because it sheds light on $\mathcal{N} = 2$ $\mathcal{T}_{1,n}(A_1)$ theory. Now, Bethe roots satisfy the following classical Bethe Ansatz equation:

$$\sum_{n=1}^{N} \frac{s_n \theta_{11}'(\mu_k - \xi_n)}{\theta_{11}(\mu_k - \xi_n)} = -\pi i \nu + \sum_{l(\neq k)} \frac{\theta_{11}'(\mu_k - \mu_l)}{\theta_{11}(\mu_k - \mu_l)}, \qquad \nu \in \text{integer}. \quad (31)$$

Regarding it as a saddle-point condition, we are led to the spectral curve analogous to (21)

$$x^2 = \left[\sum_{n=1}^{N} \frac{s_n \theta'_{11}(u-\xi_n)}{\theta_{11}(u-\xi_n)} - \sum_{k=1}^{K} \frac{\theta'_{11}(u-\mu_k)}{\theta_{11}(u-\mu_k)}\right]^2$$
$$= \sum_{n=1}^{N} s_n(s_n+1)\wp(u-\xi_n) + \sum_{n=1}^{N} H_n \zeta(u-\xi_n) + H_0$$

where

$$H_n = \sum_{i \neq n}^{N} \sum_{a=1}^{3} w_a(\xi_n - \xi_i) J_n^a J_i^a,$$

$$H_0 = \sum_{n=1}^{N} \sum_{a=1}^{3} \left\{ -\wp\left(\frac{\omega_{5-a}}{2}\right) J_n^a J_n^a \right. \tag{32}$$
$$\left. + \sum_{i \neq n} w_a(\xi_i - \xi_n)\left[\zeta\left(\xi_n - \xi_i + \frac{\omega_{5-a}}{2}\right) - \zeta\left(\frac{\omega_{5-a}}{2}\right)\right] J_n^a J_i^a \right\}.$$

Here, $\wp(u)$ and $\zeta(u)$ respectively denote Weierstrass $\wp$- and ζ-function. Periods of $\wp(u)$ are (see Appendix 4.2 for w_a)

$$\omega_1 = \omega_4 = 1, \qquad \omega_2 = \tau, \qquad \omega_3 = \tau + 1. \tag{33}$$

Notice that H_n's ($\sum H_n = 0$) are known as elliptic Gaudin Hamiltonians.[44,45] All these are elliptic counterparts of those in the rational XXX model. According to the logic of Fig. 3, it will be interesting to verify whether the XYZ one-form xdu reproduces the ϵ_1-deformed $\mathcal{N} = 2^*$ SW prepotential when $n = 1$[h]

Acknowledgments

TST thanks Kazuhiro Sakai, Hirotaka Irie and Kohei Motegi for encouragement and helpful comments. RY is supported in part by Grant-in-Aid for Scientific Research No.23540316 from Japan Ministry of Education. NY and RY are also supported in part by JSPS Bilateral Joint Projects (JSPS-RFBR collaboration).

---

[h]See also 46,47 for related discussions.

Appendix A
Definition of w_n

In this Appendix, w_n that appears in (32) will be defined according to 44,45. We choose periods of $\wp(u)$ as in (33). Weierstrass σ-function is as follows:

$$\sigma(u) = \sigma(u; \omega_1, \omega_2)$$
$$= u \prod_{\substack{(n,m) \neq (0,0) \\ n,m \in \mathbb{Z}}} \left(1 - \frac{u}{n\omega_1 + m\omega_2}\right) \quad (A.1)$$
$$\times \exp\left[\frac{u}{n\omega_1 + m\omega_2} + \frac{1}{2}\left(\frac{u}{n\omega_1 + m\omega_2}\right)^2\right].$$

Note that $\sigma(u)$ satisfies

$$\zeta(u) = \frac{\sigma'(u)}{\sigma(u)}, \qquad \wp(u) = -\zeta'(u). \quad (A.2)$$

We introduce $(e_a, \eta_a, \varsigma_a)$ which are related to ω_a by

$$e_a = \wp(\omega_a/2), \qquad \eta_a = \zeta(\omega_a/2), \qquad \varsigma_a = \sigma(\omega_a/2), \qquad a = 1, 2, 3. \quad (A.3)$$

Using them we further have

$$\sigma_{00}(u) = \frac{\exp\left[-(\eta_1 + \eta_2)u\right]}{\varsigma_3} \sigma\left(u + \frac{\omega_3}{2}\right),$$
$$\sigma_{10}(u) = \frac{\exp(-\eta_1 u)}{\varsigma_1} \sigma\left(u + \frac{\omega_1}{2}\right), \quad (A.4)$$
$$\sigma_{01}(u) = \frac{\exp(-\eta_2 u)}{\varsigma_2} \sigma\left(u + \frac{\omega_2}{2}\right).$$

Note that Jacobi's ϑ-functions are

$$\vartheta_{00}(u) = \vartheta(u; \tau) = \vartheta(u) = \sum_{n=-\infty}^{\infty} \exp\left(\pi i n^2 \tau + 2\pi i n u\right),$$
$$\vartheta_{01}(u) = \vartheta\left(u + \frac{1}{2}\right),$$
$$\vartheta_{10}(u) = \exp\left(\frac{1}{4}\pi i \tau + \pi i u\right) \vartheta\left(u + \frac{1}{2}\tau\right), \quad (A.5)$$
$$\vartheta_{11}(u) = \exp\left(\frac{1}{4}\pi i \tau + \pi i\left(u + \frac{1}{2}\right)\right) \vartheta\left(u + \frac{1}{2} + \frac{1}{2}\tau\right)$$

from which Weierstrass σ-functions are defined as below:

$$\omega_1 \exp\left(\frac{\eta_1}{\omega_1}u^2\right) \frac{\vartheta_{11}\left(\frac{u}{\omega_1}\right)}{\vartheta'_{11}(0)} = \sigma(u), \qquad (A.6)$$

$$\exp\left(\frac{\eta_1}{\omega_1}u^2\right) \frac{\vartheta_{ab}\left(\frac{u}{\omega_1}\right)}{\vartheta'_{ab}(0)} = \sigma_{ab}(u) \qquad (ab = 0).$$

Finally, $w_a(u)$ can be obtained as follows:

$$w_1(u) = \frac{\mathrm{cn}(u\sqrt{e_1-e_3}; \sqrt{\frac{e_2-e_3}{e_1-e_3}})}{\mathrm{sn}(u\sqrt{e_1-e_3}; \sqrt{\frac{e_2-e_3}{e_1-e_3}})} = \frac{\sigma_{10}(u)}{\sigma(u)} = \frac{\vartheta'_{11}(0)}{\vartheta_{10}(0)} \frac{\vartheta_{10}(u)}{\vartheta_{11}(u)},$$

$$w_2(u) = \frac{\mathrm{dn}(u\sqrt{e_1-e_3}; \sqrt{\frac{e_2-e_3}{e_1-e_3}})}{\mathrm{sn}(u\sqrt{e_1-e_3}; \sqrt{\frac{e_2-e_3}{e_1-e_3}})} = \frac{\sigma_{00}(u)}{\sigma(u)} = \frac{\vartheta'_{11}(0)}{\vartheta_{00}(0)} \frac{\vartheta_{00}(u)}{\vartheta_{11}(u)}, \qquad (A.7)$$

$$w_3(u) = \frac{1}{\mathrm{sn}(u\sqrt{e_1-e_3}; \sqrt{\frac{e_2-e_3}{e_1-e_3}})} = \frac{\sigma_{01}(u)}{\sigma(u)} = \frac{\vartheta'_{11}(0)}{\vartheta_{01}(0)} \frac{\vartheta_{01}(u)}{\vartheta_{11}(u)}.$$

References

1. N. A. Nekrasov and S. L. Shatashvili, "Quantum integrability and supersymmetric vacua," Prog. Theor. Phys. Suppl. **177** (2009) 105-119. [arXiv:0901.4748[hep-th]].
2. N. A. Nekrasov and S. L. Shatashvili, "Quantization of Integrable Systems and Four Dimensional Gauge Theories," [arXiv:0908.4052[hep-th]].
3. N. Nekrasov, A. Rosly and S. Shatashvili, "Darboux coordinates, Yang-Yang functional, and gauge theory," [arXiv:1103.3919 [hep-th]].
4. N. A. Nekrasov, "Seiberg-Witten Prepotential from Instanton Counting," Adv. Theor. Math. Phys. **7** (2004) 831-864. [hep-th/0206161].
5. N. Nekrasov and A. Okounkov, "Seiberg-Witten Theory and Random Partitions," [hep-th/0306238].
6. D. Orlando, S. Reffert, [arXiv:1011.6120 [hep-th]].
7. Y. Zenkevich, "Nekrasov prepotential with fundamental matter from the quantum spin chain," [arXiv:1103.4843 [math-ph]].
8. N. Dorey, S. Lee and T. J. Hollowood, "Quantization of Integrable Systems and a 2d/4d Duality," [arXiv:1103.5726 [hep-th]].
9. H. Y. Chen, N. Dorey, T. J. Hollowood and S. Lee,
10. D. Gaiotto and E. Witten, "Knot Invariants from Four-Dimensional Gauge Theory," [arXiv:1106.4789 [hep-th]].

11. L. F. Alday, D. Gaiotto and Y. Tachikawa, "Liouville Correlation Functions from Four-dimensional Gauge Theories," Lett. Math. Phys. **91** (2010) 167-197. [arXiv:0906.3219 [hep-th]].
12. D. Gaiotto, "N=2 dualities," [arXiv:0904.2715[hep-th]].
13. B. Feigin, E. Frenkel and N. Reshetikhin, "Gaudin model, Bethe ansatz and critical level," Comm. Math. Phys. **166** (1994) 27. [hep-th/9402022].
14. K. Ohta and T. S. Tai, JHEP **0809** (2008) 033 [arXiv:0806.2705 [hep-th]].
15. R. J. Baxter, "Partition function of the eight-vertex lattice model," Ann. Phys. **70** (1972) 193-228.
16. R. J. Baxter, "Eight vertex model in lattice statistics and one-dimensional anisotropic Heisenberg chain," Ann. Phys. **76** (1973) 1-24; 25-47; 48-71.
17. A. Gorsky, I. Krichever, A. Marshakov, A. Mironov and A. Morozov, "Integrability and Seiberg-Witten Exact Solution," Phys. Lett. **B355** (1995) 466-474. [hep-th/9505035].
18. P. C. Argyres, M. R. Plesser and A. D. Shapere, "The Coulomb phase of N=2 supersymmetric QCD," Phys. Rev. Lett. **75** (1995) 1699-1702. [hep-th/9505100].
19. R. Donagi and E. Witten, "Supersymmetric Yang-Mills Systems And Integrable Systems," Nucl. Phys. **B460** (1996) 299-334. [hep-th/9510101].
20. H. Itoyama, A. Morozov, "Integrability and Seiberg-Witten theory: Curves and periods," Nucl. Phys. **B477**, 855-877 (1996). [hep-th/9511126].
21. H. Itoyama, A. Morozov, "Prepotential and the Seiberg-Witten theory," Nucl. Phys. **B491**, 529-573 (1997). [hep-th/9512161].
22. A. Gorsky, A. Marshakov, A. Mironov and A. Morozov, "N=2 Supersymmetric QCD and Integrable Spin Chains: Rational Case $N_f < 2N_c$," Phys. Lett. **B380** (1996) 75-80. [hep-th/9603140].
23. I. M. Krichever and D. H. Phong, J. Diff. Geom. **45** (1997) 349-389. [hep-th/9604199].
24. N. Seiberg and E. Witten, "Electric-Magnetic Duality, Monopole Condensation, And Confinement In N = 2 Supersymmetric Yang-Mills Theory," Nucl. Phys. **B426** (1994) 19-52, Erratum-ibid. **B430** (1994) 485-486. [hep-th/9407087].
25. N. Seiberg and E. Witten, "Monopoles, Duality and Chiral Symmetry Breaking in N=2 Supersymmetric QCD," Nucl. Phys. **B431** (1994) 484. [arXiv:hep-th/9408099].
26. E. Witten, Nucl. Phys. **B500** (1997) 3-42. [hep-th/9703166].
27. A. Gorsky, S. Gukov and A. Mironov, "Multiscale N=2 SUSY field theories, integrable systems and their stringy/brane origin I," Nucl. Phys. **B517** (1998) 409-461. [hep-th/9707120].
28. R. Poghossian, "Deforming SW curve," JHEP **1104** (2011) 033. [arXiv:1006.4822 [hep-th]].
29. F. Fucito, J. F. Morales, D. R. Pacifici and R. Poghossian, "Gauge theories on Ω-backgrounds from non commutative Seiberg-Witten curves," JHEP **1105** (2011) 098. [arXiv:1103.4495 [hep-th]].
30. R. Dijkgraaf and C. Vafa, "Matrix Models, Topological Strings, and Supersymmetric Gauge Theories," Nucl. Phys. **B644** (2002) 3. [hep-th/0206255].

31. R. Dijkgraaf and C. Vafa, "On Geometry and Matrix Models," Nucl. Phys. **B644** (2002). [arXiv:hep-th/0207106].
32. R. Dijkgraaf and C. Vafa, "A Perturbative Window into Non-Perturbative Physics," [arXiv:[hep-th/0208048]].
33. R. Dijkgraaf and C. Vafa, "Toda Theories, Matrix Models, Topological Strings, and N=2 Gauge Systems," [arXiv:0909.2453 [hep-th]].
34. J. Teschner, "Quantization of the Hitchin moduli spaces, Liouville theory, and the geometric Langlands correspondence I," [arXiv:1005.2846 [hep-th]].
35. E. Frenkel, "Lectures on the Langlands program and conformal field theory," [hep-th/0512172]
36. T. S. Tai, "Seiberg-Witten prepotential from WZNW conformal block: Langlands duality and Selberg trace formula," arXiv:1012.4972 [hep-th].
37. O. Babelon and D. Talalaev, "On the Bethe Ansatz for the Jaynes-Cummings-Gaudin model," J. Stat. Mech. **0706** (2007) P06013. [hep-th/0703124].
38. A. B. Zamolodchikov and A. B. Zamolodchikov, "Structure constants and conformal bootstrap in Liouville field theory," Nucl. Phys. **B477** 577-605 (1996). [hep-th/9506136].
39. M. Matone, "Instantons and recursion relations in N=2 Susy gauge theory," Phys. Lett. **B357** (1995) 342-348. [hep-th/9506102].
40. J. Sonnenschein, S. Theisen and S. Yankielowicz, "On the Relation Between the Holomorphic Prepotential and the Quantum Moduli in SUSY Gauge Theories," Phys. Lett. **B367** 145-150 (1996). [hep-th/9510129].
41. T. Eguchi and S-K. Yang, "Prepotentials of N=2 Supersymmetric Gauge Theories and Soliton Equations," Mod. Phys. Lett. **A 11** 131-138 (1996). [hep-th/9510183].
42. R. Flume, F. Fucito, J. F. Morales and R. Poghossian, "Matone's Relation in the Presence of Gravitational Couplings," JHEP **0404** 008 (2004). [hep-th/0403057].
43. T. S. Tai, "Uniformization, Calogero-Moser/Heun duality and Sutherland/bubbling pants," JHEP **1010** 107 (2010). [arXiv:1008.4332 [hep-th]].
44. E. K. Sklyanin, T. Takebe, "Algebraic Bethe Ansatz for XYZ Gaudin model," Phys. Lett. **A219**, 217-225 (1996). [arXiv:q-alg/9601028].
45. E. K. Sklyanin, T. Takebe, "Separation of Variables in the Elliptic Gaudin Model," Comm. Math. Phys. **204** (1999) 17-38. [arXiv:solv-int/9807008].
46. L. F. Alday and Y. Tachikawa, Lett. Math. Phys. **94** (2010) 87 [arXiv:1005.4469 [hep-th]].
47. K. Maruyoshi and M. Taki, Nucl. Phys. B **841** (2010) 388 [arXiv:1006.4505 [hep-th]].

EXACT ANALYSIS OF CORRELATION FUNCTIONS OF THE XXZ CHAIN

TETSUO DEGUCHI[1], KOHEI MOTEGI[2] and JUN SATO[1]

[1]*Department of Physics, Graduate School of Humanities and Sciences,
Ochanomizu University 2-1-1 Ohtsuka, Bunkyo-ku, Tokyo 112-8610, Japan
E-mail:deguchi@phys.ocha.ac.jp,*

[2]*Okayama Institute for Quantum Physics,
Kyoyama 1-9-1, Okayama 700-0015, Japan
E-mail: motegi@gokutan.c.u-tokyo.ac.jp*

The correlation functions of quantum integrable models have been investigated extensively for the past twenty years. We recently studied correlation functions of the integrable higher spin XXZ chain in the massive regime. The analysis is based on the algebraic Bethe ansatz method. This article gives a simple introduction to the algebraic Bethe ansatz method, taking the spin-1/2 XXZ chain as an example. We also present the results of one point functions of integrable spin-1 XXZ chain.

Keywords: Quantum integrable system, Bethe ansatz, XXZ chain.

1. Introduction

The correlation functions of quantum integrable models have been investigated extensively for the past twenty years. Most of the researches are focused on the spin-1/2 XXZ chain, since it is one of the most fundamental integrable models, and various techniques have been developed. There are two major methods of analysis: the q-vertex operator approach[1] and the algebraic Bethe ansatz method.[2] These two methods have both advantages and disadvantages. For example, the former method can be used to analyze the dynamical structure factor in the thermodynamic limit.[3] On the other hand, the latter one can be generalized to obtain finite temperature correlation functions.[4,5]

We recently studied correlation functions of the integrable higher spin XXZ chain in the massive regime.[6] The analysis is based on the algebraic Bethe ansatz method. The method is nowadays recognized to be important because it is powerful and useful not only for analytic evaluation but also for numerical calculation. In this paper we give a simple introduction to the algebraic Bethe ansatz method,[7,8] taking the spin-1/2 XXZ chain as an example. We also briefly review the steps to calculate correlation functions by the algebraic Bethe ansatz method. We then give a preliminary report on the correlation functions of the integrable higher spin XXZ chain in the massive regime.

This paper is organized as follows. In the next section, we introduce the spin-1/2 XXZ chain. We explain the algebraic Bethe ansatz method applied to the spin-1/2 XXZ chain in section 3, and the steps to calculate correlation functions in section 4. In section 5, we give a preliminary report on correlation functions of integrable higher spin XXZ chain in the massive regime. Especially, we present the results of one point functions of integrable spin-1 XXZ chain, which were analytically evaluated from the multiple integral representation for correlation functions. Section 6 is the conclusion of this paper.

2. Spin-1/2 XXZ chain

We consider the spin-1/2 XXZ chain, whose Hamiltonian is given by

$$H = J \sum_{n=1}^{N} \{S_n^x S_{n+1}^x + S_n^y S_{n+1}^y + \Delta(S_n^z S_{n+1}^z - 1/4)\}, \tag{1}$$

where $S_n^{x,y,z} = \sigma_n^{x,y,z}/2$, $\sigma_n^{x,y,z}$ are Pauli matrices acting on the n-th site. The periodic boundary condtions are imposed: $S_{N+1}^{x,y,z} = S_1^{x,y,z}$. The Hamiltonian acts on $V_1 \otimes V_2 \otimes \cdots \otimes V_N$, where $V = \mathbf{C}^2$ is the complex two-dimensional space spanned by up and down spins. The first two terms of (1) exchange neighboring spins on the n-th and $(n+1)$-th sites, and the third term is the Ising terms. The XXZ chain is massive for $|\Delta| > 1$ and massless otherwise. When $\Delta = 1$, it is the Heisenberg XXX isotropic chain which has the $SU(2)$ symmetry. When $\Delta = 0$, it is the free-fermion model, which can be solved by Jordan-Wigner transformation. The $|\Delta| \longrightarrow \infty$ is the Ising limit.

The most straightforward method to analyze quantum spin chains is the exact diaonalization: take $|i_1, i_2 \cdots i_N\rangle (i_1, i_2 \cdots i_N = \uparrow, \downarrow)$ as the basis, write down the matrix elements of the Hamiltonian, and diagonalize the matrix to get the eigenenergies of the Hamiltonian (Fig. 1). From the data, one can see for example that the ground state is in the sector where the total spin $S^{\text{tot}} = \sum_{j=1}^{N} S_j^z$ is zero for even size.

	Spec(H/J)
$S^{tot} = 0$	$-4.08699, -3.41421, -3.25941,$
	$-3.11246, -3.11246, -2.69255, \cdots$
$S^{tot} = \pm 1$	$-3.74698, -3.05253, -3.05253,$
	$-2.88126, -2.88126, -2.59164, \cdots$
$S^{tot} = \pm 2$	$-2.81949, -2.28078, -2.28078,$
	$-2.20711, -2.20711, -1.57899, \cdots$
$S^{tot} = \pm 3$	$-1.5, -1.20711, -1.20711. -0.5,$
	$-0.5, 0.5, 0.207107, 0.207107$
$S^{tot} = \pm 4$	0

Fig. 1. Exact diagonalization of the spin-1/2 XXZ chain ($\Delta = 1/2$, $N = 8$).

However, this most general method is technically limited up to about 30 sites or so in general. On the other hand, employing the Bethe ansatz method, although limited only to quantum integrable models, we can calculate the energy of the ground state for a chain of more than 1000 sites. The exact diagonalization can be served as a check of the correctness of the Bethe ansatz. Moreover, the groundstate energy can be anlaytically evaluated in the thermodynamic limit. In the next section, we explain the algebraic Bethe ansatz method.

3. Algebraic Bethe ansatz

In this section, we develop the algebraic Bethe ansatz method for the XXZ chain. We first introduce the so-called R-matrix (Fig. 2), which acts on the tensor product space $V \otimes V$, and plays a fundamental role in quantum integrable systems and solvable lattice models:

$$R(\lambda) = \begin{pmatrix} a(\lambda) & 0 & 0 & 0 \\ 0 & b(\lambda) & c(\lambda) & 0 \\ 0 & c(\lambda) & b(\lambda) & 0 \\ 0 & 0 & 0 & a(\lambda) \end{pmatrix}, \quad (2)$$

where

$$a(\lambda) = 1, \ b(\lambda) = \frac{\sinh\lambda}{\sinh(\lambda + \eta)}, \ c(\lambda) = \frac{\sinh\eta}{\sinh(\lambda + \eta)}. \quad (3)$$

Parameter λ is called the spectral parameter, which is important for the diagonalization of the Hamiltonian and the systematic construction of conserved quantities. The parameter η is related to the anisotropy coupling constant Δ of the XXZ chain as $\Delta = \cosh\eta$. Let us denote up spin and

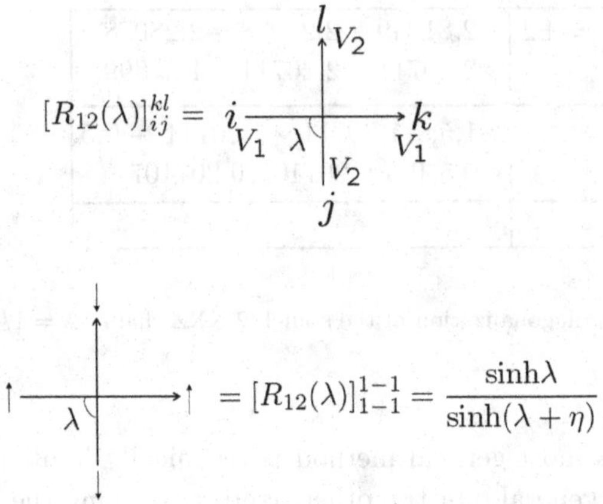

$$[R_{12}(\lambda)]_{ij}^{kl} = \begin{array}{c} {}^l\!\!\uparrow V_2 \\ i \xrightarrow[V_1]{\lambda} \xrightarrow{} k \\ {}_{V_1} \downarrow V_2 \\ j \end{array}$$

$$\begin{array}{c}\downarrow\\ \uparrow \xrightarrow{\lambda} \uparrow \\ \downarrow\end{array} = [R_{12}(\lambda)]_{1-1}^{1-1} = \frac{\sinh\lambda}{\sinh(\lambda+\eta)}$$

Fig. 2. R-matrix.

down spin by e_1 and e_{-1} respectively, and the matrix elements of the R-matrix as

$$R(\lambda)(e_i \otimes e_j) = \sum_{k,l=\pm 1} [R(\lambda)]_{ij}^{kl}(e_k \otimes e_l). \qquad (4)$$

For example, the matrix element for $i = 1, j = -1, k = 1$ and $l = -1$ is given by

$$[R(\lambda)]_{1-1}^{1-1} = b(\lambda) = \frac{\sinh\lambda}{\sinh(\lambda+\eta)}. \qquad (5)$$

The R-matrix satisfies the Yang-Baxter relation (Fig. 3)

$$R_{12}(\lambda-\mu)R_{13}(\lambda)R_{23}(\mu) = R_{23}(\mu)R_{13}(\lambda)R_{12}(\lambda-\mu), \qquad (6)$$

acting on $V_1 \otimes V_2 \otimes V_3$. Here, the symbol $R_{13}(\lambda)$ means that the R-matrix acts on the tensor product of the first and third spaces. The Yang-Baxter relation consists of 64 equations. One can check each equation by direct

$$R_{12}(\lambda - \mu)R_{13}(\lambda)R_{23}(\mu) = R_{23}(\mu)R_{13}(\lambda)R_{12}(\lambda - \mu)$$

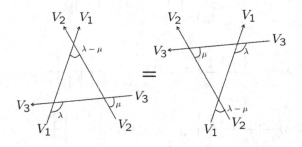

Fig. 3. Yang-Baxter relation.

calculation. For example,

$$[R_{12}(\lambda - \mu)R_{13}(\lambda)R_{23}(\mu)]_{-11-1}^{-1-11} - [R_{23}(\mu)R_{13}(\lambda)R_{12}(\lambda - \mu)]_{-11-1}^{-1-11}$$
$$=[R(\lambda - \mu)]_{-1-1}^{-1-1}[R(\lambda)]_{-11}^{-11}[R(\mu)]_{1-1}^{-11}$$
$$- [R(\mu)]_{1-1}^{-11}[R(\lambda)]_{-1-1}^{-1-1}[R(\lambda - \mu)]_{-11}^{-11} - [R(\mu)]_{-11}^{-11}[R(\lambda)]_{1-1}^{-11}[R(\lambda - \mu)]_{-11}^{1-1}$$
$$=\frac{\sinh\lambda\sinh\eta}{\sinh(\lambda + \eta)\sinh(\mu + \eta)}$$
$$- \frac{\sinh\eta\sinh(\lambda - \mu)}{\sinh(\mu + \eta)\sinh(\lambda - \mu + \eta)} - \frac{\sinh^2\eta\sinh\mu}{\sinh(\mu + \eta)\sinh(\lambda + \eta)\sinh(\lambda - \mu + \eta)}$$
$$=0. \qquad (7)$$

The monodromy matrix $T_a(\lambda)$ acting on the tensor product space $V_a \otimes (V_1 \otimes \cdots \otimes V_N)$ is constructed from the R-matrices as (Fig. 4)

$$T_a(\lambda) = R_{aN}(\lambda - \eta/2)\cdots R_{a1}(\lambda - \eta/2) \qquad (8)$$
$$= \begin{pmatrix} A(\lambda) & B(\lambda) \\ C(\lambda) & D(\lambda) \end{pmatrix}, \qquad (9)$$

where V_a is referred to as the 'auxiliary' space, while $V_1, \cdots, V_N$ are called the 'quantum' spaces. The elements of the monodromy matrix are the A, B, C, D operators. Note that the B operator has the role of creating a down spin.

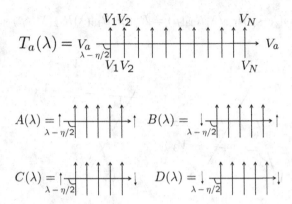

Fig. 4. Monodromy matrix.

Utilizing the Yang-Baxter relation repeatedly, one has the RTT relation (Fig. 5)

$$R_{ab}(\lambda - \mu)T_a(\lambda)T_b(\mu) = T_b(\mu)T_a(\lambda)R_{ab}(\lambda - \mu), \qquad (10)$$

from which the commutativity of the transfer matrix $\mathcal{T}(\lambda) = \text{tr}_a T_a(\lambda)$ with

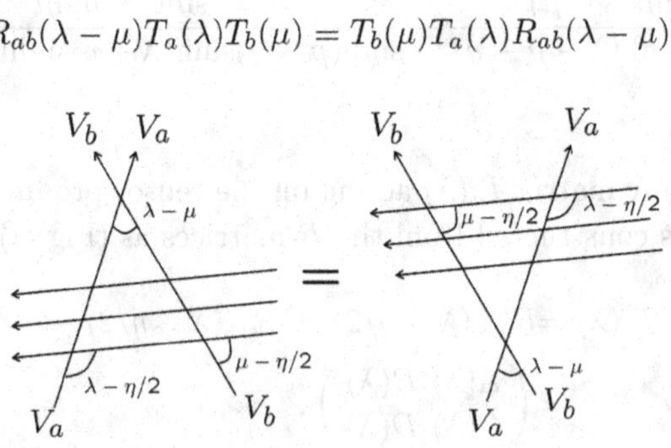

$$R_{ab}(\lambda - \mu)T_a(\lambda)T_b(\mu) = T_b(\mu)T_a(\lambda)R_{ab}(\lambda - \mu)$$

Fig. 5. RTT relation.

two different values of the spectral parameter follows

$$[\mathcal{T}(\lambda), \mathcal{T}(\mu)] = 0. \tag{11}$$

The Hamiltonian of the spin-1/2 XXZ chain is obtained as the logarithmic derivative of the transfer matrix

$$H = \frac{J}{2}\sinh\eta \frac{\mathrm{d}}{\mathrm{d}\lambda}\ln\mathcal{T}(\lambda)|_{\lambda=\eta/2}, \ \Delta = \cosh\eta. \tag{12}$$

Instead of directly diagonalizing the Hamiltonian, we first diagonalize the transfer matrix by the algebraic Bethe ansatz method. The eigenvalues of the Hamiltonian are obtained after that.

Let us now diagonalize the transfer matrix. We prepare the 'vacuum', i.e., the state with all spins up
$|0\rangle = |\uparrow_1\uparrow_2 \cdots \uparrow_N\rangle \in V_1 \otimes V_2 \otimes \cdots \otimes V_N$. Note that the B operator creates a down spin, and acts on $V_1 \otimes V_2 \otimes \cdots \otimes V_N$. Then the vector

$$\prod_{k=1}^{r} B(\lambda_k)|0\rangle, \tag{13}$$

belongs to the space $V_1 \otimes V_2 \otimes \cdots \otimes V_N$ with r down-spins and $(N-r)$ up-spins.

We need some more preparations. Note the following relations hold

$$A(\lambda)B(\mu) = \frac{1}{b(\mu-\lambda)}B(\mu)A(\lambda) - \frac{c(\mu-\lambda)}{b(\mu-\lambda)}B(\lambda)A(\mu), \tag{14}$$

$$D(\lambda)B(\mu) = \frac{1}{b(\lambda-\mu)}B(\mu)D(\lambda) - \frac{c(\lambda-\mu)}{b(\lambda-\mu)}B(\lambda)D(\mu), \tag{15}$$

$$[B(\lambda), B(\mu)] = 0, \tag{16}$$

$$A(\lambda)|0\rangle = |0\rangle, \tag{17}$$

$$D(\lambda)|0\rangle = b(\lambda - \eta/2)^N|0\rangle. \tag{18}$$

The first three relations are elements of the RTT relation (Figs. 6 and 7). The last two ones can be checked directly. First, utilizing (14) and (16),

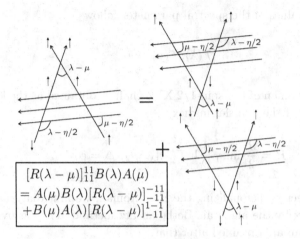

Fig. 6. An element of the RTT relation (14).

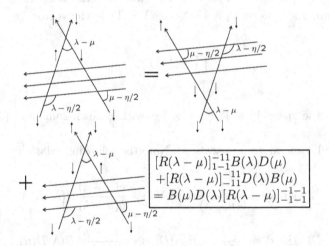

Fig. 7. An element of the RTT relation (15).

one can show the following relation by induction

$$A(\lambda)\prod_{k=1}^{r}B(\lambda_k) = \prod_{k=1}^{r}\frac{1}{b(\lambda_k-\lambda)}\prod_{k=1}^{r}B(\lambda_k)A(\lambda)$$
$$-\sum_{j=1}^{r}\frac{c(\lambda_j-\lambda)}{b(\lambda_j-\lambda)}\prod_{k\neq j}\frac{1}{b(\lambda_k-\lambda_j)}B(\lambda)\prod_{k\neq j}^{r}B(\lambda_k)A(\lambda_j).$$
(19)

From (15) and (16), in the same way, we get

$$D(\lambda) \prod_{k=1}^{r} B(\lambda_k)|0\rangle = \prod_{k=1}^{r} \frac{1}{b(\lambda - \lambda_k)} \prod_{k=1}^{r} B(\lambda_k) D(\lambda)$$
$$- \sum_{j=1}^{r} \frac{c(\lambda - \lambda_j)}{b(\lambda - \lambda_j)} \prod_{k \neq j} \frac{1}{b(\lambda_j - \lambda_k)} B(\lambda) \prod_{k \neq j} B(\lambda_k) D(\lambda_j). \tag{20}$$

In (19) and (20), we moved the A and D operators to the right of B operators. Since the actions of A and D operators on the vacuum give the vacuum with eigenvalues as factors (17) and (18), we see that the actions of A and D operators on the state $\prod_{k=1}^{r} B(\lambda_k)|0\rangle$ are expressed in terms only of B operators. Thus, combining (17) and (19), one has

$$A(\lambda) \prod_{k=1}^{r} B(\lambda_k)|0\rangle = \prod_{k=1}^{r} \frac{1}{b(\lambda_k - \lambda)} \prod_{k=1}^{r} B(\lambda_k)|0\rangle$$
$$- \sum_{j=1}^{r} \frac{c(\lambda_j - \lambda)}{b(\lambda_j - \lambda)} \prod_{k \neq j} \frac{1}{b(\lambda_k - \lambda_j)} B(\lambda) \prod_{k \neq j} B(\lambda_k)|0\rangle, \tag{21}$$

and from (18) and (20),

$$D(\lambda) \prod_{k=1}^{r} B(\lambda_k)|0\rangle$$
$$= b(\lambda - \eta/2)^N \prod_{k=1}^{r} \frac{1}{b(\lambda - \lambda_k)} \prod_{k=1}^{r} B(\lambda_k)|0\rangle$$
$$- \sum_{j=1}^{r} b(\lambda_j - \eta/2)^N \frac{c(\lambda - \lambda_j)}{b(\lambda - \lambda_j)} \prod_{k \neq j} \frac{1}{b(\lambda_j - \lambda_k)} B(\lambda) \prod_{k \neq j} B(\lambda_k)|0\rangle. \tag{22}$$

Combining these two relations, we get

$$(A(\lambda) + D(\lambda)) \prod_{k=1}^{r} B(\lambda_k)|0\rangle$$
$$= \left\{ \prod_{k=1}^{r} \frac{1}{b(\lambda_k - \lambda)} + b(\lambda - \eta/2)^N \prod_{k=1}^{r} \frac{1}{b(\lambda - \lambda_k)} \right\} \prod_{k=1}^{r} B(\lambda_k)|0\rangle$$
$$- \sum_{j=1}^{r} \left\{ \frac{c(\lambda_j - \lambda)}{b(\lambda_j - \lambda)} \prod_{k \neq j} \frac{1}{b(\lambda_k - \lambda_j)} \right.$$
$$\left. + b(\lambda_j - \eta/2)^N \frac{c(\lambda - \lambda_j)}{b(\lambda - \lambda_j)} \prod_{k \neq j} \frac{1}{b(\lambda_j - \lambda_k)} \right\} B(\lambda) \prod_{k \neq j} B(\lambda_k)|0\rangle. \tag{23}$$

The "**Bethe ansatz**" is the requirement that "**let $\prod_{k=1}^{r} B(\lambda_k)|0\rangle$ be an eigenstate of the transfer matrix**". The first term of the rhs of (23) is the 'wanted term', while the other ones are 'unwanted terms'. In order for $\prod_{k=1}^{r} B(\lambda_k)|0\rangle$ to be an eigenstate, all coefficients of the 'unwanted terms' have to be zero. The conditions exactly lead to the Bethe ansatz equations:

$$\left\{\frac{\sinh(\lambda_k - \eta/2)}{\sinh(\lambda_k + \eta/2)}\right\}^N = \prod_{\substack{j=1 \\ j\neq k}}^{r} \frac{\sinh(\lambda_k - \lambda_j - \eta)}{\sinh(\lambda_k - \lambda_j + \eta)}, \qquad (24)$$

for $k = 1, 2, \cdots, r$. Then from the eigenvalue of the transfer matrix

$$(A(\lambda) + D(\lambda))\prod_{k=1}^{r} B(\lambda_k)|0\rangle = \tau(\lambda)\prod_{k=1}^{r} B(\lambda_k)|0\rangle, \qquad (25)$$

$$\tau(\lambda) = \prod_{j=1}^{r}\frac{\sinh(\lambda_j - \lambda - \eta)}{\sinh(\lambda_j - \lambda)} + \left\{\frac{\sinh(\lambda - \eta/2)}{\sinh(\lambda + \eta/2)}\right\}^N \prod_{j=1}^{r} \frac{\sinh(\lambda_j - \lambda + \eta)}{\sinh(\lambda_j - \lambda)}, \qquad (26)$$

one gets the eigenvalue of the Hamiltonian as

$$E/J = \frac{1}{2}\sinh\eta \frac{d}{d\lambda}\ln\tau(\lambda)|_{\lambda=\eta/2} = \sum_{j=1}^{r}\frac{\sinh^2\eta}{\cosh(2\lambda_j) - \cosh\eta}. \qquad (27)$$

Taking the logarithm of the Bethe ansatz equations (24), one has

$$\lambda_k = \text{arctanh}\left[i\tanh(\eta/2)\tan\left\{\frac{\pi I_k}{N} + \frac{1}{N}\sum_{j=1}^{r}\arctan\left(\frac{-i\tanh(\lambda_k - \lambda_j)}{\tanh\eta}\right)\right\}\right], \qquad (28)$$

for $k = 1, 2, \cdots, r$. Here, $\{I_k\}$ are given by either all integers or all half-integers, which characterize the eigenstate. For example, when N even, the ground state corresponds to $r = N/2$ (half-filling) and the set of integers or half-integers $\{I_k\} = \{-N/4 + 1/2, -N/4 + 3/2, \ldots, N/4 - 1/2\}$.

One can use the equations (28) for numerical evaluation of the Bethe ansatz roots. For example, for $\Delta = 1/2$ and $N = 8$, the set of I_k's $\{I_k\} = \{-3/2, -1/2, 1/2, 3/2\}$ characterizes the ground state and we have $\{\lambda_k\} = \{-0.5469, -0.135285, 0.135285, 0.5469\}$. Inserting this data into (27), one gets $E/J = -4.08699$, which coincides with the result of the exact diagonalization. One should stress that contrary to the exact diagonalization method, this procedure works for chains with more than 1,000 or even 10,000 sites. These data are nowadays used for evaluation of correlation functions like dynamical structure factors,[9-11] for example.

There are many variants of the Bethe ansatz besides the algebraic Bethe ansatz. The first one developed by Bethe himself is called the coordinate Bethe ansatz, which seems to be the most natural idea. In this method, the wavefunction is assumed to be written as a linear combination of plane waves, and the Bethe ansatz equations appear by imposing the periodic boundary conditions. There are also several types of the Bethe ansatz such as Baxter's Q-operator method, the analytic Bethe ansatz and so on. However, for computing correlation functions, one needs the eigenstate, which can be constructed by neither the Q-operator method nor the analytic Bethe ansatz. Also we need a good formula for calculating the inner product between a given eigenstate and the dual of the eigenstate. It is only available for the algebraic Bethe ansatz at the moment. This is the reason why we have explained the algebraic Bethe ansatz. In the next section, we briefly review the steps to calculate correlation functions by the algebraic Bethe ansatz.

4. Steps to calculate correlation functions

The procedure to calculate form factors and correlation functions by the algebraic Bethe ansatz method was established for the spin-1/2 XXZ chain at the end of 20-th century[2]. It consists of the following steps.

Step 1. Formulate form factors and correlation functions in the quantum inverse scattering language

First, one has to express form factors and correlation functions in terms of the $ABCD$ operators to begin calculating them by the algebraic Bethe ansatz. What we should do essentially is to express local operators such as σ_2^+ in terms of the global $ABCD$ operators. To do this we introduce inhomogeneous parameters $\xi_1, \xi_2, \ldots, \xi_N$, in quantum spaces. By this change, the monodromy matrix includes inhomogeneous parameters

$$T_a(\lambda) = R_{aN}(\lambda - \xi_N) \cdots R_{a1}(\lambda - \xi_1) \tag{29}$$

$$= \begin{pmatrix} A(\lambda) & B(\lambda) \\ C(\lambda) & D(\lambda) \end{pmatrix}. \tag{30}$$

The XXZ chain can be recovered by taking the homogeneous limit $\xi_j = \eta/2$ ($j = 1, \cdots, N$).

Introducing these inhomogeneous parameters, one can express local operators in the quantum inverse scattering language

$$|1/2\rangle_{ii}\langle 1/2| = \prod_{\alpha=1}^{i-1}(A+D)(\xi_\alpha)A(\xi_i)\prod_{\alpha=i+1}^{N}(A+D)(\xi_\alpha), \qquad (31)$$

$$|-1/2\rangle_{ii}\langle 1/2| = \prod_{\alpha=1}^{i-1}(A+D)(\xi_\alpha)B(\xi_i)\prod_{\alpha=i+1}^{N}(A+D)(\xi_\alpha), \qquad (32)$$

$$|1/2\rangle_{ii}\langle -1/2| = \prod_{\alpha=1}^{i-1}(A+D)(\xi_\alpha)C(\xi_i)\prod_{\alpha=i+1}^{N}(A+D)(\xi_\alpha), \qquad (33)$$

$$|-1/2\rangle_{ii}\langle -1/2| = \prod_{\alpha=1}^{i-1}(A+D)(\xi_\alpha)D(\xi_i)\prod_{\alpha=i+1}^{N}(A+D)(\xi_\alpha). \qquad (34)$$

Using these formulas, it is easy to formulate form factors and correlation functions.

Step 2. Deform the formulated form factors and correlation functions into forms such that the scalar product formula can be applied

The next thing to do is to deform the formulated form factors and correlation functions into forms to which one can apply the scalar product formula. This process can be accomplished by acting the $ABCD$ operators repeatedly on the dual of the Bethe state $\langle 0| \prod_{k=1}^{r} C(\lambda_k)$. The RTT relation and the action of the A and D operators on the vacuum are the basic ingredients for this step.

Step 3. Apply the scalar product formula to get the determinant representation of form factors and the multiple sum representation of correlation functions

After deforming the formulated form factors and correlation functions into appropriate forms, you now apply the scalar product formula, which is also called Slavnov formula. The statement is as follows: if $\{\mu\}$ or $\{\lambda\}$ is a set of Bethe roots, we have

$$\langle 0| \prod_{j=1}^{r} C(\mu_j) \prod_{k=1}^{r} B(\lambda_k)|0\rangle = \frac{\det T(\{\mu_j\}, \{\lambda_k\})}{\det V(\{\mu_j\}, \{\lambda_k\})}, \quad (35)$$

where matrices T and V are given by

$$T_{ab} = \frac{\partial}{\partial \lambda_a} \tau(\mu_b, \{\lambda_k\}), \quad V_{ab} = \frac{1}{\sinh(\mu_b - \lambda_a)}. \quad (36)$$

Applying this scalar product formula, one obtains the determinant representation of form factors and the multiple sum representation of correlation functions.

Step 4. Take the thermodynamic limit to get the multiple integral representation for correlation functions

The last step is to take the thermodynamic limit to turn the sums into integrals. Since the Bethe roots of the ground state are real, the integral runs from $-\infty$ to ∞ in the massless regime, and from $-\pi/2$ to $\pi/2$ in the massive regime for the zero-magnetic field. Information about the Bethe roots of the ground state is encoded in the density function of the Bethe roots, which can be analytically evaluated by solving the Lieb equation. As a result of this step, one finally gets the mutliple integral representation for correlation functions.

5. Integrable higher spin XXZ chain

Recently, by the use of the algebraic Bethe ansatz which we reviewed in the previous sections, we studied the correlation functions of the integrable higher spin XXZ chain. For example, we calculated the one point functions of the integrable spin-1 XXZ chain, whose Hamiltonian is given by

$$H_{spin-1} = J \sum_{j=1}^{N} \Big\{ \vec{S}_j \cdot \vec{S}_{j+1} - (\vec{S}_j \cdot \vec{S}_{j+1})^2$$

$$- \frac{1}{2}(q - q^{-1})^2 [S_j^z S_{j+1}^z - (S_j^z S_{j+1}^z)^2 + 2(S_j^z)^2]$$

$$- (q + q^{-1} - 2)[(S_j^x S_{j+1}^x + S_j^y S_{j+1}^y) S_j^z S_{j+1}^z + S_j^z S_{j+1}^z (S_j^x S_{j+1}^x + S_j^y S_{j+1}^y)] \Big\}, \quad (37)$$

where $q = \exp(-\eta)$. For example, we find the probability $P(1)$ that the state of spin is $+1$, has the following double integral representation

$$P(1) = \frac{1}{4\pi^2} \prod_{n=1}^{\infty} \left(\frac{1-q^{2n}}{1+q^{2n}}\right)^4 \int_{-\pi/2}^{\pi/2} d\nu_1 \int_{-\pi/2}^{\pi/2} d\nu_2$$

$$\times \left\{ \frac{\sin(\nu_1 + i\eta/2)\sin(\nu_2 - i\eta/2)}{\sin i\eta \sin(\nu_2 - \nu_1 - i\epsilon)} \frac{\theta_3(\nu_1/\pi; \tau)\theta_3(\nu_2/\pi; \tau)}{\theta_4(\nu_1/\pi; \tau)\theta_4(\nu_2/\pi; \tau)} \right.$$

$$\left. - \frac{\sin(\nu_1 - i\eta/2)\sin(\nu_2 + i\eta/2)}{\sin i\eta \sin(\nu_2 - \nu_1 + 2i\eta)} \frac{\theta_3(\nu_1/\pi; \tau)\theta_3(\nu_2/\pi; \tau)}{\theta_4(\nu_1/\pi; \tau)\theta_4(\nu_2/\pi; \tau)} \right\}. \quad (38)$$

Here, the theta functions

$$\theta_3(v; \tau) = \sum_{n=-\infty}^{\infty} e^{\pi i \tau n^2 + 2\pi i n v}, \quad (39)$$

$$\theta_4(v; \tau) = \sum_{n=-\infty}^{\infty} e^{\pi i \tau n^2 + 2\pi i n (v+1/2)}, \quad (40)$$

with period 1 and quasiperiod $\tau = i\eta/\pi$ appear in the integrand. This double integral is analytically evaluated as

$$P(1) = \frac{q + q^{-1}}{q^{-1} - q} \sum_{n=-\infty}^{\infty} \frac{1}{(q^n + q^{-n})^2} - \frac{2}{(q - q^{-1})^2}, \quad (41)$$

and agrees with the result of numerical exact diagonalization. One can also show $P(0)$, which is the probability that the state of spin is 0, is represented by utilizing the following double intergal

$$P(0) = 1 + \frac{1}{2\pi^2} \prod_{n=1}^{\infty} \left(\frac{1-q^{2n}}{1+q^{2n}}\right)^4 \int_{-\pi/2}^{\pi/2} d\nu_1 \int_{-\pi/2}^{\pi/2} d\nu_2$$

$$\times \left\{ \frac{\sin(\nu_1 + i\eta/2)\sin(\nu_2 - i\eta/2)}{\sin i\eta \sin(\nu_1 - \nu_2 + i\epsilon)} \frac{\theta_3(\nu_1/\pi; \tau)\theta_3(\nu_2/\pi; \tau)}{\theta_4(\nu_1/\pi; \tau)\theta_4(\nu_2/\pi; \tau)} \right.$$

$$\left. - \frac{\sin(\nu_1 - i\eta/2)\sin(\nu_2 + i\eta/2)}{\sin i\eta \sin(\nu_1 - \nu_2 - 2i\eta)} \frac{\theta_3(\nu_1/\pi; \tau)\theta_3(\nu_2/\pi; \tau)}{\theta_4(\nu_1/\pi; \tau)\theta_4(\nu_2/\pi; \tau)} \right\}, \quad (42)$$

and can be evaluated as

$$P(0) = \frac{2(q+q^{-1})}{q-q^{-1}} \sum_{n=-\infty}^{\infty} \frac{1}{(q^n+q^{-n})^2} + \left(\frac{q+q^{-1}}{q-q^{-1}}\right)^2. \qquad (43)$$

The expressions of the one point functions (41), (43) can also be compared with the partial results obtained from the q-vertex operator approach. Details will be given in the forthcoming paper.[6]

6. Conclusion

In this paper, we have formulated an introduction to the algebraic Bethe ansatz method, taking spin-1/2 XXZ chain as an example. As we mentioned in the introduction, the algebraic Bethe ansatz method is gaining its importance since it allows us to analytically evaluate correlation functions, even for the integrable higher spin XXZ chain about which we have presented some results on the model in the last section. Although it is only possible for some particular integrable models, the algebraic Bethe ansatz method is also efficient for numerical calculation of correlation functions, and one can analyze much larger systems than the exact diagonalization method.

Acknowledgments

We would like to thank H. Konno, A. Nishino, K. Sakai, T. Tai and R. Weston for discussions and comments.

Appendix A: Evaluation of (42)

In this Appendix, we evaluate the double integrals in (42) to get (43). The following double integrals appear in (42)

$$\frac{1}{2\pi^2} \prod_{n=1}^{\infty} \left(\frac{1-q^{2n}}{1+q^{2n}}\right)^4 \int_{-\pi/2}^{\pi/2} d\nu_1 \int_{-\pi/2}^{\pi/2} d\nu_2$$

$$\times \left\{ \frac{\sin(\nu_1+i\eta/2)\sin(\nu_2-i\eta/2)}{\sin i\eta \sin(\nu_1-\nu_2+i\epsilon)} \frac{\theta_3(\nu_1/\pi;\tau)\theta_3(\nu_2/\pi;\tau)}{\theta_4(\nu_1/\pi;\tau)\theta_4(\nu_2/\pi;\tau)} \right.$$

$$\left. - \frac{\sin(\nu_1-i\eta/2)\sin(\nu_2+i\eta/2)}{\sin i\eta \sin(\nu_1-\nu_2-2i\eta)} \frac{\theta_3(\nu_1/\pi;\tau)\theta_3(\nu_2/\pi;\tau)}{\theta_4(\nu_1/\pi;\tau)\theta_4(\nu_2/\pi;\tau)} \right\}. \qquad (A.1)$$

We consider the first term in (A.1), and change the integration contour in the ν_1 plane as

$$\int_{-\pi/2}^{\pi/2} d\nu_1 \int_{-\pi/2}^{\pi/2} d\nu_2 \frac{\sin(\nu_1 + i\eta/2)\sin(\nu_2 - i\eta/2)}{\sin i\eta \sin(\nu_1 - \nu_2 + i\epsilon)} \frac{\theta_3(\nu_1/\pi;\tau)\theta_3(\nu_2/\pi;\tau)}{\theta_4(\nu_1/\pi;\tau)\theta_4(\nu_2/\pi;\tau)}$$

$$= \int_{-\pi/2}^{\pi/2} d\nu_2 \frac{\sin(\nu_2 - i\eta/2)\theta_3(\nu_2/\pi;\tau)}{\sin i\eta \, \theta_4(\nu_2/\pi;\tau)} \int_{-\pi/2}^{\pi/2} d\nu_1 \frac{\sin(\nu_1 + i\eta/2)\theta_3(\nu_1/\pi;\tau)}{\sin(\nu_1 - \nu_2 + i\epsilon)\theta_4(\nu_1/\pi;\tau)}$$

$$= \int_{-\pi/2}^{\pi/2} d\nu_2 \frac{\sin(\nu_2 - i\eta/2)\theta_3(\nu_2/\pi;\tau)}{\sin i\eta \, \theta_4(\nu_2/\pi;\tau)}$$

$$\times \left\{ \left(\int_{-\pi/2}^{-\pi/2-i\eta} + \int_{-\pi/2-i\eta}^{\pi/2-i\eta} + \int_{\pi/2-i\eta}^{\pi/2} \right) d\nu_1 \right.$$

$$\left. - 2\pi i (\text{Res}_{\nu_1=-i\eta/2} + \text{Res}_{\nu_1=\nu_2-i\epsilon}) \right\} \times \frac{\sin(\nu_1 + i\eta/2)\theta_3(\nu_1/\pi;\tau)}{\sin(\nu_1 - \nu_2 + i\epsilon)\theta_4(\nu_1/\pi;\tau)}$$

$$= \int_{-\pi/2}^{\pi/2} d\nu_2 \frac{\sin(\nu_2 - i\eta/2)\theta_3(\nu_2/\pi;\tau)}{\sin i\eta \, \theta_4(\nu_2/\pi;\tau)}$$

$$\times \left\{ -\int_{-\pi/2}^{\pi/2} d\nu_1 \frac{\sin(\nu_1 - i\eta/2)\theta_3(\nu_1/\pi;\tau)}{\sin(\nu_1 - \nu_2 - i\eta)\theta_4(\nu_1/\pi;\tau)} \right.$$

$$\left. - 2\pi i \frac{\sin(\nu_2 + i\eta/2)\theta_3(\nu_2/\pi;\tau)}{\theta_4(\nu_2/\pi;\tau)} \right\}$$

$$= \int_{-\pi/2}^{\pi/2} d\nu_1 \int_{-\pi/2}^{\pi/2} d\nu_2 \frac{\sin(\nu_1 - i\eta/2)\sin(\nu_2 - i\eta/2)}{\sin i\eta \sin(\nu_2 - \nu_1 + i\eta)} \frac{\theta_3(\nu_1/\pi;\tau)\theta_3(\nu_2/\pi;\tau)}{\theta_4(\nu_1/\pi;\tau)\theta_4(\nu_2/\pi;\tau)}$$

$$- 2\pi i \int_{-\pi/2}^{\pi/2} d\nu_2 \frac{\sin(\nu_2 + i\eta/2)\sin(\nu_2 - i\eta/2)}{\sin i\eta} \frac{\theta_3^2(\nu_2/\pi;\tau)}{\theta_4^2(\nu_2/\pi;\tau)}. \quad (A.2)$$

Next, one changes the contour of integration of the first term in (A.2) with respect to ν_2 as

$$\int_{-\pi/2}^{\pi/2} d\nu_1 \int_{-\pi/2}^{\pi/2} d\nu_2 \frac{\sin(\nu_1 - i\eta/2)\sin(\nu_2 - i\eta/2)}{\sin i\eta \sin(\nu_2 - \nu_1 + i\eta)} \frac{\theta_3(\nu_1/\pi;\tau)\theta_3(\nu_2/\pi;\tau)}{\theta_4(\nu_1/\pi;\tau)\theta_4(\nu_2/\pi;\tau)}$$

$$= \int_{-\pi/2}^{\pi/2} d\nu_1 \frac{\sin(\nu_1 - i\eta/2)\theta_3(\nu_1/\pi;\tau)}{\sin i\eta \, \theta_4(\nu_1/\pi;\tau)} \int_{-\pi/2}^{\pi/2} d\nu_2 \frac{\sin(\nu_1 - i\eta/2)\theta_3(\nu_2/\pi;\tau)}{\sin(\nu_2 - \nu_1 + i\eta)\theta_4(\nu_2/\pi;\tau)}$$

$$= \int_{-\pi/2}^{\pi/2} d\nu_1 \frac{\sin(\nu_1 - i\eta/2)\theta_3(\nu_1/\pi;\tau)}{\sin i\eta \theta_4(\nu_1/\pi;\tau)}$$

$$\times \left\{ \left(\int_{-\pi/2}^{-\pi/2+i\eta} + \int_{-\pi/2+i\eta}^{\pi/2+i\eta} + \int_{\pi/2+i\eta}^{\pi/2} \right) d\nu_2 + 2\pi i \text{Res}_{\nu_2 = i\eta/2} \right\}$$

$$\times \frac{\sin(\nu_1 - i\eta/2)\theta_3(\nu_2/\pi;\tau)}{\sin(\nu_2 - \nu_1 + i\eta)\theta_4(\nu_2/\pi;\tau)}$$

$$= \int_{-\pi/2}^{\pi/2} d\nu_1 \frac{\sin(\nu_1 - i\eta/2)\theta_3(\nu_1/\pi;\tau)}{\sin i\eta \theta_4(\nu_1/\pi;\tau)} \int_{-\pi/2}^{\pi/2} d\nu_2 \frac{\sin(\nu_2 + i\eta/2)\theta_3(\nu_2/\pi;\tau)}{\sin(\nu_1 - \nu_2 - 2i\eta)\theta_4(\nu_2/\pi;\tau)}$$

$$= \int_{-\pi/2}^{\pi/2} d\nu_1 \int_{-\pi/2}^{\pi/2} d\nu_2 \frac{\sin(\nu_1 - i\eta/2)\sin(\nu_2 + i\eta/2)}{\sin i\eta \sin(\nu_1 - \nu_2 - 2i\eta)} \frac{\theta_3(\nu_1/\pi;\tau)\theta_3(\nu_2/\pi;\tau)}{\theta_4(\nu_1/\pi;\tau)\theta_4(\nu_2/\pi;\tau)},$$
(A.3)

which exactly cancels the second term in (A.1), i.e., inserting (A.2) and (A.3), the double integrals (A.1) reduce to a single integral

$$\frac{1}{2\pi^2} \prod_{n=1}^{\infty} \left(\frac{1-q^{2n}}{1+q^{2n}} \right)^4 \int_{-\pi/2}^{\pi/2} d\nu_1 \int_{-\pi/2}^{\pi/2} d\nu_2$$

$$\times \left\{ \frac{\sin(\nu_1 + i\eta/2)\sin(\nu_2 - i\eta/2)}{\sin i\eta \sin(\nu_1 - \nu_2 + i\epsilon)} \frac{\theta_3(\nu_1/\pi;\tau)\theta_3(\nu_2/\pi;\tau)}{\theta_4(\nu_1/\pi;\tau)\theta_4(\nu_2/\pi;\tau)} \right.$$

$$\left. - \frac{\sin(\nu_1 - i\eta/2)\sin(\nu_2 + i\eta/2)}{\sin i\eta \sin(\nu_1 - \nu_2 - 2i\eta)} \frac{\theta_3(\nu_1/\pi;\tau)\theta_3(\nu_2/\pi;\tau)}{\theta_4(\nu_1/\pi;\tau)\theta_4(\nu_2/\pi;\tau)} \right\}$$

$$= -\frac{1}{\pi \text{sh}\eta} \prod_{n=1}^{\infty} \left(\frac{1-q^{2n}}{1+q^{2n}} \right)^4 \int_{-\pi/2}^{\pi/2} d\nu_2 \frac{\sin(\nu_2 + i\eta/2)\sin(\nu_2 - i\eta/2)\theta_3^2(\nu_2/\pi;\tau)}{\theta_4^2(\nu_2/\pi;\tau)}.$$
(A.4)

This single integral (A.4) can be evaluated by applying the addition formula for the trigonometric functions and utilizing

$$\int_{-\pi/2}^{\pi/2} dx \frac{\cos 2x \theta_3^2(x/\pi;\tau)}{\theta_4^2(x/\pi;\tau)} = \frac{4\pi \text{sh}\eta}{\text{ch}2\eta - 1} \left(\frac{1+q^{2n}}{1-q^{2n}} \right)^4, \quad (A.5)$$

$$\prod_{n=1}^{\infty} \left(\frac{1-q^{2n}}{1+q^{2n}} \right)^2 \frac{\theta_3(x/\pi;\tau)}{\theta_4(x/\pi;\tau)} = \sum_{n=-\infty}^{\infty} \frac{e^{2ixn}}{\text{ch} n\eta}, \quad (A.6)$$

as

$$-\frac{1}{\pi\text{sh}\eta}\prod_{n=1}^{\infty}\left(\frac{1-q^{2n}}{1+q^{2n}}\right)^4 \int_{-\pi/2}^{\pi/2} d\nu_2 \frac{\sin(\nu_2+i\eta/2)\sin(\nu_2-i\eta/2)\theta_3^2(\nu_2/\pi;\tau)}{\theta_4^2(\nu_2/\pi;\tau)}$$

$$=-\frac{1}{2\pi\text{sh}\eta}\prod_{n=1}^{\infty}\left(\frac{1-q^{2n}}{1+q^{2n}}\right)^4$$

$$\times \left(\text{ch}\eta \int_{-\pi/2}^{\pi/2} d\nu_2 \frac{\theta_3^2(\nu_2/\pi;\tau)}{\theta_4^2(\nu_2/\pi;\tau)} - \int_{-\pi/2}^{\pi/2} d\nu_2 \cos 2\nu_2 \frac{\theta_3^2(\nu_2/\pi;\tau)}{\theta_4^2(\nu_2/\pi;\tau)}\right)$$

$$=-\frac{\text{ch}\eta}{2\text{sh}\eta}\sum_{n=-\infty}^{\infty}\frac{1}{\text{ch}^2 n\eta} + \frac{2}{\text{ch}2\eta-1}. \qquad (A.7)$$

From (42), (A.4) and (A.7), one gets

$$P(0) = \frac{\text{ch}2\eta+1}{\text{ch}2\eta-1} - \frac{\text{ch}\eta}{2\text{sh}\eta}\sum_{n=-\infty}^{\infty}\frac{1}{\text{ch}^2 n\eta}. \qquad (A.8)$$

changing the parameter from η to $q = \exp(-\eta)$, we have (43).

References

1. M. Jimbo, K. Miki, T. Miwa and A. Nakayashiki, *Phys. Lett. A* **168**, (1992) 256.
2. N. Kitanine, J.M. Maillet and V. Terras, *Nucl. Phys. B* **567**, (2000) 554.
3. J.S. Caux, H. Konno, M. Sorrell and R. Weston, *Phys. Rev. Lett.* **106**, (2011) 217203.
4. F. Göhmann, A. Klümper and A. Seel, *J. Phys. A* **37**, (2004) 7625.
5. K. Sakai, *J. Phys. A* **40**, (2007) 7523.
6. T. Deguchi, K. Motegi and J. Sato, paper in preparation.
7. L.A. Takhtajan and L.D. Faddeev, *Russ. Math. Surveys* **34** (1979) 11.
8. V.E. Korepin, N.M. Bogoliubov and A.G. Izergin, *Quantum Inverse Scattering Method and Correlation functions* (Cambridge University Press, 1993).
9. D. Biegel, M. Karbach and G. Müller, *Europhys. Lett.* **59**, (2002) 882.
10. J. Sato, M. Shiroishi and M. Takahashi, *J. Phys. Soc. Jpn.* **73**, (2004) 3008.
11. J.S. Caux, R. Hagemans and J.M. Maillet, *J. Stat. Mech.* (2005) P09003.

CLASSICAL ANALOGUE OF WEAK VALUE IN STOCHASTIC PROCESS

H. TOMITA

Research Center of Quantum Computing, Kinki University
Higashi-Osaka, 577-8502, Japan
E-mail: tomita@alice.math.kindai.ac.jp

One of the remarkable notions in the recent development of quantum physics is the weak value related to weak measurements. We emulate it as a two-time conditional expectation in a classical stochastic model. We use the well known symmetrized form of the master equation, which is formally equivalent to the wave equation in quantum mechanics apart from the fact that wave functions are always real. The origin of the unusual behaviors of the weak value such as the negative probability and the abnormal enhancement of some expectations becomes clearer in the present case, where the two-time conditional probability has no ambiguity of imaginary/complex values.

Keywords: Weak value, Stochastic process, Two-time conditional probability, Stochastic Ising model.

1. Introduction

The weak value is a derived notion of the weak measurement proposed by Aharonov et al.,[1] which has brought a new understanding of quantum observations. The weak measurement[2] means that it hardly disturbs the quantum superposed state when it is performed with large uncertainty. The reason of the strange nature of this quantum measurement is that the weak value is defined as an expectation over strongly restricted paths with the condition of the post-selected final state.

Suppose we have started from a pre-selected initial state at $t = t_\mathrm{i}$. If we measure some observable Q at t ($> t_\mathrm{i}$), the wave function $\Psi(t)$ collapses to one of the eigenstates of Q. And we can find only the probability distribution $|\Psi(t)|^2$, not the probability amplitude $\Psi(t)$, by repeating the measurement and adopting all of the observed data. If we discard the main data by restricting the paths to the post-selected ones only, we can expect to get some informations on the state $\Psi(t)$ before collapsing, because the post-selected paths are described with the same propagator as the forward evolution of the pre-selected paths by changing only the sign of the time t according to the time-reversal symmetry of quantum mechanics.

The weak value of an observable Q with a given initial state $|i\rangle$ at $t = t_i$ and a final state $|f\rangle$ at $t = t_f$ is defined by[3]

$$\langle Q \rangle^{\text{w}}_{(\text{f;i})} = \frac{\langle f|e^{-i(t_f-t)H} Q e^{-i(t-t_i)H}|i\rangle}{\langle f|e^{-i(t_f-t_i)H}|i\rangle}, \quad (t_i \leq t \leq t_f) \tag{1}$$

where H is the Hamiltonian and the unit $\hbar = 1$ is used. This quantity is related to a *weak measurement* as follows: Let us introduce a *meter* to measure the observable Q of the target system at $t = t_0$ by a weak interaction

$$H_{\text{int}}(t) = g\delta(t - t_0) Q \otimes p,$$

where p is the momentum operator of the probe of the meter and g is a small coupling constant. Suppose the initial state of the meter, $\varphi(x)$ in the coordinate representation of the probe position x has a sufficiently broad uncertainty Δ, i.e. a variance Δ^2. It can be easily shown that for the restricted paths from i to f, the meter state for $t > t_0$ is given by

$$\langle f|e^{-i(t_f-t)H} e^{-igQ\otimes p} e^{-i(t-t_i)H}|i\rangle \varphi(x)$$
$$\simeq \langle f|e^{-i(t_f-t_i)H}|i\rangle \varphi\left(x - g\langle Q\rangle^{\text{w}}_{(\text{f;i})}\right), \tag{2}$$

where $p = -i\partial/\partial x$ is used. Note that the weak value defined by Eq.(1) is complex in general. Then the shift of the expectation of the probe position x is given by the real part of the weak value, $g\text{Re}[\langle Q\rangle^{\text{w}}_{(\text{f;i})}]$, while the shift of the expectation of the momentum p is found to be equal to the imaginary part, $(g/2\Delta^2)\text{Im}[\langle Q\rangle^{\text{w}}_{(\text{f;i})}]$ by using a Fourier transformation.

An early interpretation of the quantity defined by Eq.(1) is the probability *amplitude*[4] which yields a pre- and post-selected, two-time conditional probability (TTCP). When it is applied to a projection operator $|q\rangle\langle q|$ onto one of the eigenstates of an observable Q, one finds

$$|\langle|q\rangle\langle q|\rangle^{\text{w}}_{(\text{f;i})}|^2 = \left|\frac{\langle f|e^{-i(t_f-t)H}|q\rangle\langle q|e^{-i(t-t_i)H}|i\rangle}{\langle f|e^{-i(t_f-t_i)H}|i\rangle}\right|^2 = \frac{P(f|q)P(q|i)}{P(f|i)}. \tag{3}$$

The last expression is to be shown equal to $P(q|f \cap i)$ in Sec.3 by using Bayes identities. (See the footnote c in Sec.3.) Here $P(*|C)$ is the standard notation for a conditional probability with a condition C (or a transition probability from the state C to $*$), e.g.

$$P(f|q) = |\langle f|e^{-i(t_f-t)H}|q\rangle|^2, \quad \text{etc.}$$

Another interpretation, somewhat formal one, is the TTCP itself.[3,5] Let us rewrite the usual quantum expectation of $Q(t) = e^{i(t-t_i)H}Qe^{-i(t-t_i)H}$ with respect to the state $|i\rangle$ in the following identical form,

$$\langle i|Q(t)|i\rangle = \sum_f \langle i|e^{i(t_f-t_i)H}|f\rangle\langle f|e^{-i(t_f-t)H}Qe^{-i(t-t_i)H}|i\rangle$$
$$= \sum_f \langle Q\rangle^{\text{w}}_{(f;i)} |\langle f|e^{-i(t_f-t_i)H}|i\rangle|^2, \qquad (4)$$

where the last factor of Eq.(4) is $P(f|i)$. This may yield an interpretation of the weak value as a complex, raw stochastic variable. Nevertheless, if it is applied to $|q\rangle\langle q|$ again, it reads as a conditional probability equation,

$$P(q|i) = |\langle q|e^{-i(t-t_i)H}|i\rangle|^2 = \sum_f \langle |q\rangle\langle q|\rangle^{\text{w}}_{(f;i)} P(f|i). \qquad (5)$$

In addition, we have a sum-rule,

$$\sum_q \langle |q\rangle\langle q|\rangle^{\text{w}}_{(f;i)} = 1,$$

because of the completeness, $\sum_q |q\rangle\langle q| = I$ (= identity). Therefore, if we remind a type of the Bayes statistics relations, [a]

$$P(q|i) = \sum_f P(f \cap q|i) = \sum_f P(q|f \cap i)P(f|i),$$

the weak value of the projection operator $|q\rangle\langle q|$, though it may be *complex*, can be interpreted formally as a TTCP itself with a couple of pre- and post-selections, i and f. Further, the weak value of an operator $Q = \sum_q q|q\rangle\langle q|$, or $A = \sum_q a(q)|q\rangle\langle q|$ in general, can be interpreted as the two-time conditional expectations (TTCE) of them with respect to this virtual TTCP.

Because of this rather fictitious interpretation, the virtual conditional probability happens to be *negative*[6] and it causes an abnormal enhancement of the weak value of some observables greater than their inherent norms.[5] These strange behaviors are closely related. That is, at least if a probability set $\{P(q)\}$ is real, it can be expected that we have a partial sum satisfying

$$\sum_{P(q)\geq 0} P(q) = 1 - \sum_{P(q)<0} P(q) > 1, \qquad (6)$$

whenever there exists a negative part. This is the essential reason of a possibility of the unusual enhancement of some expectations.[6,7]

---

[a] $P(f \cap q|i) = P(q \cap f \cap i)/P(i) = P(q|f \cap i)P(f \cap i)/P(i) = P(q|f \cap i)P(f|i)$
For the Bayes identity, see the equation (∗) in the footnote c in Sec.3.

The purpose of the present work is to emulate these strange behaviors clearer by using a classical stochastic model, in which we can avoid the ambiguity of the complex probability in the above quantum problem.[8] We survey a conventional transformation of the stochastic master equation to a self-adjoint form in the following section. A good analogy with the quantum mechanics is found by applying it to the TTCP. This is shown in Sec.3. An example of the stochastic Ising model which shows an abnormal enhancement of the expectations of some quantities with respect to TTCP is given in Sec.4. In Sec.5 we discuss an extension of TTCP to a density matrix form to complete the analogy with the quantum mechanics. The last section is devoted to brief summary and discussions.

2. Self-adjoint form of stochastic master equation

First let us survey the well-known transformation[9] to a self-adjoint form of the stochastic master equation.

Let $\boldsymbol{x}$ be a set of stochastic variable(s) described by a time-dependent conditional probability, $P(\boldsymbol{x}, t|\boldsymbol{x}_i, t_i)$ for $t \geq t_i$, which obeys the following stationary, Markovian master equation, i.e. the Chapman-Kolmogorov *forward* equation,

$$\frac{\partial}{\partial t} P(\boldsymbol{x}, t|\boldsymbol{x}_i, t_i) = -\sum_{\boldsymbol{x}'} W(\boldsymbol{x} \to \boldsymbol{x}') P(\boldsymbol{x}, t|\boldsymbol{x}_i, t_i)$$
$$+ \sum_{\boldsymbol{x}'} W(\boldsymbol{x}' \to \boldsymbol{x}) P(\boldsymbol{x}', t|\boldsymbol{x}_i, t_i)$$
$$= -\sum_{\boldsymbol{x}'} L(\boldsymbol{x}, \boldsymbol{x}') P(\boldsymbol{x}', t|\boldsymbol{x}_i, t_i), \qquad (7)$$

where

$$L(\boldsymbol{x}, \boldsymbol{x}') = \delta(\boldsymbol{x} - \boldsymbol{x}') \sum_{\boldsymbol{x}''} W(\boldsymbol{x} \to \boldsymbol{x}'') - W(\boldsymbol{x}' \to \boldsymbol{x}).$$

The matrix L has an eigenvalue $\lambda_0 = 0$ corresponding to the steady state,

$$P_0(\boldsymbol{x}) = \lim_{t-t_i \to \infty} P(\boldsymbol{x}, t|\boldsymbol{x}_i, t_i).$$

Let us introduce a *wave function* related to this forward conditional probability by

$$\psi(\boldsymbol{x}, t|\boldsymbol{x}_i, t_i) = \phi_0(\boldsymbol{x})^{-1} P(\boldsymbol{x}, t|\boldsymbol{x}_i, t_i), \quad (t \geq t_i) \qquad (8)$$

where $\phi_0(x) = P_0(x)^{1/2}$. This function ψ obeys the forward wave equation,

$$\frac{\partial}{\partial t}\psi(x,t) = -\sum_{x'} H(x,x')\psi(x',t), \qquad (9)$$

where H is defined by

$$H(x,x') = \phi_0(x)^{-1} L(x,x') \phi_0(x'). \qquad (10)$$

For the time being the initial condition (x_i, t_i) in ψ is abbreviated. The function $\phi_0(x)$ is an eigenfunction of Eq.(9) for $\lambda_0 = 0$.

The merit of this transformation is that the eigenvalue problem of a given master equation is simplified, if the matrix H is symmetric, i.e.

$$H(x,x') = H(x',x).$$

This situation is widely expected when the detailed balance condition, i.e. the time-reversal symmetry,[10]

$$P_0(x)W(x \to x') = P_0(x')W(x' \to x),$$

or equivalently,

$$L(x,x')P_0(x') = L(x',x)P_0(x), \qquad (11)$$

is satisfied. [b] In this case the eigenvalues of H are all real, and non-negative, if the steady state is stable. Therefore, $\phi_0(x)$ is the ground state.

A useful example is the Fokker-Planck equation for a single, continuous stochastic variable x,

$$\frac{\partial}{\partial t}P(x,t) = -\mathcal{L}[x]P(x,t), \quad \mathcal{L}[x] = -\frac{\partial}{\partial x}\left(F'(x) + \frac{\epsilon}{2}\frac{\partial}{\partial x}\right), \qquad (12)$$

which describes a one-dimensional Brownian motion in a potential $F(x)$ with a small diffusion constant ϵ. By using its steady state solutions,

$$P_0(x) \propto \exp[-2F(x)/\epsilon] \quad \text{and} \quad \phi_0(x) \propto \exp[-F(x)/\epsilon],$$

we find the continuous variable version of the above formulations,

$$\mathcal{H}[x] = \frac{1}{\epsilon}\left[-\frac{\epsilon^2}{2}\frac{\partial^2}{\partial x^2} + V(x)\right], \quad V(x) = \frac{1}{2}\left[F'(x)^2 - \epsilon F''(x)\right]. \qquad (13)$$

---

[b]It should be noted that the time reversal symmetry is assumed in this form of the probability flow and not on the transition probability itself, the latter being satisfied in quantum mechanics. This is the reason why we need the above transformation to obtain a self-adjoint formulation like quantum mechanics.

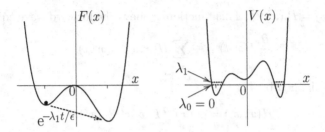

Fig. 1. Stochastic decay process of the metastable state.

Thus the Fokker-Planck equation is transformed into a self-adjoint form of an imaginary-time Schrödinger equation,

$$-\epsilon \frac{\partial}{\partial t}\psi(x,t) = \left[-\frac{\epsilon^2}{2}\frac{\partial^2}{\partial x^2} + V(x)\right]\psi(x,t),$$

and its eigenvalue problem results in a familiar one of the quantum mechanics.

Figure.1 shows an early application[11] to the so-called Kramers escape problem. The stochastic decay (or escape) rate of the metastable state in a double-well potential $F(x)$ is given by the first excited eigenvalue λ_1 of the corresponding Schrödinger potential $V(x)$. The first excited state is almost degenerate with the ground state for a small diffusion constant ϵ.

3. Two-time conditional probability

So far the quantum mechanical reformulation merely helps us to simplify the eigenvalue problem of a given master equation. None of remarkable quantum mechanical phenomena appears, until we are concerned with the TTCP,

$$P(\boldsymbol{x},t|\boldsymbol{x}_\text{f},t_\text{f};\boldsymbol{x}_\text{i},t_\text{i}), \quad t_\text{i} \leq t \leq t_\text{f}\,. \quad (\,;\text{denoting 'and', or } \cap\,) \qquad (14)$$

By using the Markovian property and the well-known relation between joint and conditional probabilities repeatedly, [c] the TTCP can be written in the

---

[c]By using the primitive identity of the Bayes theorem,

$$P(A|B)P(B) = P(B|A)P(A) = P(A \cap B), \qquad (*)$$

we find in abbreviated notations,

$$P(x|f \cap i) = \frac{P(f \cap x \cap i)}{P(f \cap i)} = \frac{P(f|x \cap i)P(x \cap i)}{P(f \cap i)} = \frac{P(f|x)P(x|i)P(i)}{P(f \cap i)},$$

following form with a pair of wave functions as

$$P(\bm{x},t|\bm{x}_\mathrm{f},t_\mathrm{f};\bm{x}_\mathrm{i},t_\mathrm{i}) = \frac{1}{\langle\psi_\mathrm{f}|\psi_\mathrm{i}\rangle}\overline{\psi}(\bm{x},t|\bm{x}_\mathrm{f},t_\mathrm{f})\psi(\bm{x},t|\bm{x}_\mathrm{i},t_\mathrm{i}), \qquad (15)$$

where the associated wave function denoted by $\overline{\psi}$ is related to the so-called *posterior* conditional probability, $\overline{P}(\bm{x},t|\bm{x}_\mathrm{f},t_\mathrm{f})$ for $t \leq t_\mathrm{f}$, by

$$\overline{\psi}(\bm{x},t|\bm{x}_\mathrm{f},t_\mathrm{f}) = \phi_0(\bm{x})^{-1}\overline{P}(\bm{x},t|\bm{x}_\mathrm{f},t_\mathrm{f}), \qquad (16)$$

and obeys the *backward* wave equation,

$$\frac{\partial}{\partial t}\overline{\psi}(\bm{x},t) = \sum_{\bm{x}'} H^\dagger(\bm{x},\bm{x}')\overline{\psi}(\bm{x}',t). \qquad (17)$$

Here $H^\dagger$ is the hermite conjugate of H, i.e. the transposed matrix in the present case. The eigensystem is common with the forward equation Eq.(9), when H is hermitian, i.e. real and symmetric as has been assumed here.

The denominator in Eq.(15) is the weight of overlap between the two wave functions defined by an inner product,

$$\langle\psi_\mathrm{f}|\psi_\mathrm{i}\rangle = \sum_{\bm{x}} \overline{\psi}(\bm{x},t|\bm{x}_\mathrm{f},t_\mathrm{f})\psi(\bm{x},t|\bm{x}_\mathrm{i},t_\mathrm{i}). \qquad (18)$$

Of course this quantity is real, while the corresponding quantity in the quantum mechanics is complex in general.

Let us define the ket- and the bra-vectors by

$$|\psi_\mathrm{i}(t)\rangle = \{\psi(\bm{x},t|\bm{x}_\mathrm{i},t_\mathrm{i})\}^T \quad \text{and} \quad \langle\psi_\mathrm{f}(t)| = \{\overline{\psi}(\bm{x},t|\bm{x}_\mathrm{f},t_\mathrm{f})\}. \qquad (19)$$

Then the wave equations (9) and (17) by assuming $H^\dagger = H$ are rewritten in the quantum mechanical form as

$$\frac{\partial}{\partial t}|\psi_\mathrm{i}(t)\rangle = -H|\psi_\mathrm{i}(t)\rangle \quad \text{and} \quad \frac{\partial}{\partial t}\langle\psi_\mathrm{f}(t)| = \langle\psi_\mathrm{f}(t)|H, \qquad (20)$$

respectively. Henceforth, H is called the Hamitonian.

---

where the Markovness, i.e. $P(f|x \cap i) = P(f|x)$ is assumed for the time order $t_\mathrm{f} \geq t \geq t_\mathrm{i}$. By applying the identity (∗) to $P(f|x)$ again, we obtain a symmetric expression,

$$P(x|f \cap i) = \frac{P(x|f)P(f)}{P(x)}\frac{P(x|i)P(i)}{P(f \cap i)} = \frac{1}{R(f,i)}\frac{P(x|f)P(x|i)}{P(x)},$$

where the first denominator $R(f,i)$ is given by

$$R(f,i) = \frac{P(f \cap i)}{P(f)P(i)} = \sum_x \frac{P(x|f)P(x|i)}{P(x)},$$

because of the normalization condition, $\sum_x P(x|f \cap i) = 1 \; \forall f \cap i$.

By using this pair of the Schrödinger equations it is shown that the overlap integral, or the inner product $\langle\psi_f|\psi_i\rangle$ given by Eq.(18) does not depend on the current time t, i.e.

$$\frac{\partial}{\partial t}\langle\psi_f|\psi_i\rangle = \langle\psi_f(t)|H|\psi_i(t)\rangle - \langle\psi_f(t)|H|\psi_i(t)\rangle = 0.$$

It should be noted that the present wave function ψ satisfies a conservation law only in this meaning coupled with its adjoint $\bar\psi$ as Eq.(18). In addition, it can be shown that this overlap integral has the following properties in the respective limits;

(i) $\lim_{t_f - t_i \to \infty} \langle\psi_f|\psi_i\rangle = 1,$

(ii) $\lim_{t_f - t_i \to 0} \langle\psi_f|\psi_i\rangle = [\phi_0(\boldsymbol{x}_f)\phi_0(\boldsymbol{x}_i)]^{-1}\delta(\boldsymbol{x}_f - \boldsymbol{x}_i).$ (21)

Note that the TTCE (two-time conditional expectation) of a physical quantity Q with respect to TTCP defined by

$$\langle Q\rangle^w_{(f;i)} = \sum_{\boldsymbol{x}} Q(\boldsymbol{x})P(\boldsymbol{x},t|\boldsymbol{x}_f,t_f;\boldsymbol{x}_i,t_i) = \frac{\langle\psi_f(t)|Q|\psi_i(t)\rangle}{\langle\psi_f|\psi_i\rangle}, \quad (22)$$

has just the analogous form of the weak value in the quantum mechanics.

Thus the TTCP is a nonlinear quantity composed of a product of a pair of the forward and the backward wave functions, and cannot be described by a closed, linear evolution equation. Then it happens that the principle of the probability superposition is violated and the interference of wave functions may occur. However, its example is omitted here because none of remarkable phenomena from this view point has been found, yet. The reason may be that the wave functions are always *real* and positive in the present case. Therefore, let us discuss only the weak value in the rest.

4. Stochastic model of classical Ising spins

An example is a pair of the classical Ising spin $\sigma = \pm 1$ having an exchange interaction,

$$E(\boldsymbol{x}) = -J\sigma_1\sigma_2,$$

where $\boldsymbol{x} = (\sigma_1, \sigma_2)$. Let us number the stochastic variable $\boldsymbol{x}$ in the order, $(1,1)$, $(1,-1)$, $(-1,1)$, $(-1,-1)$ and choose the following transition matrices,

$$W = \begin{pmatrix} 0 & 1 & 1 & 0 \\ p^2 & 0 & 0 & p^2 \\ p^2 & 0 & 0 & p^2 \\ 0 & 1 & 1 & 0 \end{pmatrix} \text{ or } L = \begin{pmatrix} 2p^2 & -1 & -1 & 0 \\ -p^2 & 2 & 0 & -p^2 \\ -p^2 & 0 & 2 & -p^2 \\ 0 & -1 & -1 & 2p^2 \end{pmatrix}, \quad (23)$$

where $p = e^{-\beta J}$, $\beta = 1/k_B T$. Evidently this transition matrix W satisfies the detailed balance condition,

$$e^{-\beta E(\boldsymbol{x})} W(\boldsymbol{x} \to \boldsymbol{x}') = e^{-\beta E(\boldsymbol{x}')} W(\boldsymbol{x}' \to \boldsymbol{x}),$$

at the steady state, i.e. the thermal equilibrium of a temperature T. With the use of the equilibrium distribution function,

$$P_0(\boldsymbol{x}) = \frac{1}{2(1+p^2)} (1,\ p^2,\ p^2,\ 1) \quad \text{and} \quad \phi_0(\boldsymbol{x}) = \frac{1}{\sqrt{2(1+p^2)}} (1,\ p,\ p,\ 1),$$

we find the corresponding hermitian Hamiltonian,

$$H = \begin{pmatrix} 2p^2 & -p & -p & 0 \\ -p & 2 & 0 & -p \\ -p & 0 & 2 & -p \\ 0 & -p & -p & 2p^2 \end{pmatrix}$$

$$= (1+p^2)\,\sigma_0 \otimes \sigma_0 - (1-p^2)\,\sigma_z \otimes \sigma_z - p\,(\sigma_0 \otimes \sigma_x + \sigma_x \otimes \sigma_0), \quad (24)$$

where σ_x and σ_z are the usual Pauli matrices and σ_0 denotes the two dimensional unit matrix I_2. This is the Hamiltonian of a pair of *quantum* Ising spins with an exchange interaction in a transverse magnetic field.

The eigenvalues and the eigenstates of this Hamiltonian H,

$$\begin{cases} \lambda_0 = 0,\ \lambda_1 = 2p^2,\ \lambda_2 = 2,\ \lambda_3 = 2(1+p^2), \\ |0\rangle = \frac{1}{\sqrt{2(1+p^2)}} [\,|\uparrow\uparrow\rangle + p\,|\uparrow\downarrow\rangle + p\,|\downarrow\uparrow\rangle + |\downarrow\downarrow\rangle\,], \\ |1\rangle = \frac{1}{\sqrt{2}} [\,|\uparrow\uparrow\rangle - |\downarrow\downarrow\rangle\,], \\ |2\rangle = \frac{1}{\sqrt{2}} [\,|\uparrow\downarrow\rangle - |\downarrow\uparrow\rangle\,], \\ |3\rangle = \frac{1}{\sqrt{2(1+p^2)}} [\,p\,|\uparrow\uparrow\rangle - |\uparrow\downarrow\rangle - |\downarrow\uparrow\rangle + p\,|\downarrow\downarrow\rangle\,], \end{cases} \quad (25)$$

can be easily obtained, where $|0\rangle = |\phi_0\rangle$, the ground state. Here the familiar notations $\uparrow, \downarrow$ are used for $\sigma = \pm 1$. Note that the first excited state is almost degenerate with the ground state for a small transition probability p^2.

By using this eigensystem we can calculate the state vectors, $|\psi_i(t)\rangle$ and $\langle\psi_f(t)|$ for arbitrary initial and final states in just the same manner of the elementary quantum mechanics except for the fact that the time t is imaginary.

Strange behaviors can be expected only when the paths from i to f are very rare cases, because the post-selection causes little effect when the

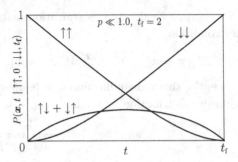

Fig. 2. Two-time conditional probability.

paths are dominant ones. Then let us consider the case where the initial and the final states differ from each other. Let

$$x_i = \uparrow\uparrow \text{ at } t = 0 \text{ and } x_f = \downarrow\downarrow \text{ at } t = t_f,$$

that is,

$$P(x, 0) = (1, 0, 0, 0) \text{ and } \overline{P}(x, t_f) = (0, 0, 0, 1),$$

or equivalently,

$$|\psi_i(0)\rangle = \sqrt{2(1 + p^2)} \, |\uparrow\uparrow\rangle \text{ and } \langle\psi_f(t_f)| = \sqrt{2(1 + p^2)} \, \langle\downarrow\downarrow|.$$

By using the eigenvector expansion we obtain,

$$|\psi_i(t)\rangle = |0\rangle + \sqrt{1 + p^2} \, e^{-\lambda_1 t} \, |1\rangle + p \, e^{-\lambda_3 t} \, |3\rangle,$$
$$\langle\psi_f(t)| = \langle 0| - \sqrt{1 + p^2} \, e^{-\lambda_1 (t_f - t)} \langle 1| + p \, e^{-\lambda_3 (t_f - t)} \langle 3|,$$
(26)

and

$$\langle\psi_f|\psi_i\rangle = 1 - (1 + p^2) \, e^{-\lambda_1 t_f} + p^2 \, e^{-\lambda_3 t_f} \quad (> 0). \tag{27}$$

The TTCP is shown in Figure.2. This result itself is very natural and well-expected, all probabilities being always non-negative.

A strange behavior appears when we use the basis $\{|k\rangle, \ k = 0, 1, 2, 3\}$, the eigenstates of the Hamiltonian H instead of the spin states $\{|x\rangle = |\sigma_1 \sigma_2\rangle\}$. We can calculate the virtual probability, i.e. the TTCE of the projection operator $|k\rangle\langle k|$ onto each eigenstate $|k\rangle$ in the same manner. The result is given by

$$P(0,t) = \frac{\langle\psi_f(t)|0\rangle\langle 0|\psi_i(t)\rangle}{\langle\psi_f|\psi_i\rangle} = \frac{1}{\langle\psi_f|\psi_i\rangle},$$

$$P(1,t) = \frac{\langle\psi_f(t)|1\rangle\langle 1|\psi_i(t)\rangle}{\langle\psi_f|\psi_i\rangle} = -\frac{(1+p^2)e^{-\lambda_1 t_f}}{\langle\psi_f|\psi_i\rangle} \quad (<0),$$

$$P(2,t) = \frac{\langle\psi_f(t)|2\rangle\langle 2|\psi_i(t)\rangle}{\langle\psi_f|\psi_i\rangle} = 0,$$

$$P(3,t) = \frac{\langle\psi_f(t)|3\rangle\langle 3|\psi_i(t)\rangle}{\langle\psi_f|\psi_i\rangle} = \frac{p^2 e^{-\lambda_3 t_f}}{\langle\psi_f|\psi_i\rangle}.$$

(28)

The fictitious negative probability is found in $P(1,t)$. Of course the completeness of the probability, $\sum_{k=0}^{3} P(k,t) = 1$, is satisfied evidently because of Eq.(27).

This negativity is precisely expected from the signs of the expansion coefficients of the eigenvectors in the right side of Eq.(25). Some of inner products $\langle x|k\rangle$ between two basis systems $\{|x\rangle = |\sigma_1\sigma_2\rangle\}$ and $\{|k\rangle\}$ are found to be negative. Then some part of the virtual TTCP happens to be negative, when the initial and the final states differ from each other.

On the contrary, when the both states are the same, this situation cannot be expected, because the negative inner products, if any, would be squared. For example, when we select as

$$x_i = x_f = \uparrow\uparrow \text{ at } t=0 \text{ and } t=t_f,$$

we find the corresponding virtual probabilities all positive, i.e. [d]

$$P(0,t) = \frac{\langle\psi_i(t)|0\rangle\langle 0|\psi_i(t)\rangle}{\langle\psi_i|\psi_i\rangle} = \frac{1}{\langle\psi_i|\psi_i\rangle},$$

$$P(1,t) = \frac{\langle\psi_i(t)|1\rangle\langle 1|\psi_i(t)\rangle}{\langle\psi_i|\psi_i\rangle} = \frac{(1+p^2)e^{-\lambda_1 t_f}}{\langle\psi_i|\psi_i\rangle},$$

$$P(2,t) = \frac{\langle\psi_i(t)|2\rangle\langle 2|\psi_i(t)\rangle}{\langle\psi_i|\psi_i\rangle} = 0,$$

$$P(3,t) = \frac{\langle\psi_i(t)|3\rangle\langle 3|\psi_i(t)\rangle}{\langle\psi_i|\psi_i\rangle} = \frac{p^2 e^{-\lambda_3 t_f}}{\langle\psi_i|\psi_i\rangle},$$

(29)

where

$$\langle\psi_i|\psi_i\rangle = 1 + (1+p^2)\,e^{-\lambda_1 t_f} + p^2\,e^{-\lambda_3 t_f}.$$

---

[d] Note that the bra- and the ket-vectors in these expressions denote

$$|\psi_i(t)\rangle = e^{-tH}|\psi_i\rangle \text{ and } \langle\psi_i(t)| = \langle\psi_i|e^{-(t_f-t)H},$$

from the present definitions Eq.(19) of them, and the overlap integral $\langle\psi_i|\psi_i\rangle$ in the denominators is better to be written explicitly as $\langle\psi_i(t)|\psi_i(t)\rangle$, or $\langle\psi_i|e^{-t_f H}|\psi_i\rangle$ to avoid a confusion, as if $\langle\psi_i|\psi_i\rangle = 1$ in the usual quantum mechanical notation.

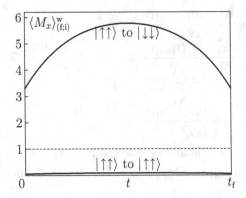

Fig. 3. Abnormal and normal TTCE of the transverse magnetization M_x for the transition probability $p = 0.2$ and $p^2 t_f = 0.01$. The indicators $|\uparrow\uparrow\rangle$ and $|\downarrow\downarrow\rangle$ denote the initial and the final states of the respective TTCP.

This is a general conclusion for two different bases $\{e_i\}$ and $\{e'_j\}$ of any *real* vector space, because at least one of the inner products, $\{e_i \cdot e'_j\}$ must be negative.

That is, the negative probability can be expected at least when

(1) the initial and the final states differ from each other,
(2) and the orthogonal basis of the intermediate projection differs from the basis of the initial and the final selections, so that one of the inner products is negative.

The same situation may occur in the quantum system between different sets of eigenvectors of *non-commutative* observables, say, P, Q. When we select the initial and the final states as different eigenstates of P, the virtual TTCP, i.e. the weak value of the projection operator $|q\rangle\langle q|$ onto some of the eigenstates of Q can be negative. This setting is sufficient for the condition[5] to find the strange weak value.

A strange behavior related to this negative probability is the abnormal enhancement of some observables as is stated in Sec.1. An example is shown in Fig.3 for a quantity, say, the *transverse magnetization*,

$$M_x = \frac{1}{2}(\sigma_x \otimes \sigma_0 + \sigma_0 \otimes \sigma_x). \tag{30}$$

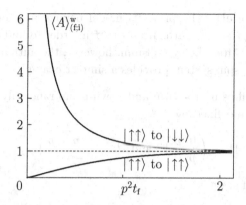

Fig. 4. Abnormal and normal TTCE of $A = \sigma_x \otimes \sigma_x$ for $p = 0.2$.

An abnormal behavior

$$\langle M_x \rangle^{\text{w}}_{(\text{f;i})} = \frac{1}{\langle \psi_\text{f} | \psi_\text{i} \rangle} \left[\frac{2p}{1+p^2} \left(1 - p^2 e^{-\lambda_3 t_\text{f}}\right) - \frac{1-p^2}{1+p^2} \left(e^{-\lambda_3 t} + e^{-\lambda_3 (t_\text{f}-t)}\right) \right]$$
$$> 1, \tag{31}$$

is found for sufficiently small p and t_f. Note that the natural norm of M_x must be less than 1, because the eigenvalue spectrum of M_x is $\{-1, 0, 0, 1\}$. When the transition rate is very small, i.e. $p^2 t_\text{f} \ll 1$, we find

$$\langle M_x \rangle^{\text{w}}_{(\text{f;i})} \gg 1.$$

A plain reason of this singular behavior is that the overlap integral $\langle \psi_\text{f} | \psi_\text{i} \rangle$ in the denominator may be expected to be very small owing to (ii) of Eq.(21), whenever the initial and the final states differ from each other, i.e. $\boldsymbol{x}_\text{i} \neq \boldsymbol{x}_\text{f}$. This means that to reach $\boldsymbol{x}_\text{f} = (\downarrow\downarrow)$ starting from $\boldsymbol{x}_\text{i} = (\uparrow\uparrow)$ in a given time occurs scarcely and is far from the main flow of the conditional probability. On the contrary none of such strange behaviors are found when $\boldsymbol{x}_\text{i} = \boldsymbol{x}_\text{f}$, e.g. $\boldsymbol{x}_\text{i} = \boldsymbol{x}_\text{f} = (\uparrow\uparrow)$. The result for the latter case for the same parameters as the upper abnormal case is shown by the lower curve in Fig.3, its maximum being ~ 0.09 at $t = t_\text{f}/2$ and minimum ~ 0.05 at $t = 0$ and t_f.

In Fig.4 the TTCE of another quantity $A = \sigma_x \otimes \sigma_x$ having a spectrum $\{-1, -1, 1, 1\}$ are shown also. Note that A is commutative with H and is a conserved quantity. Then the horizontal axis in this figure shows a parameter of the transition probability instead of the current time itself.

(Single spin model) The previous model having an energy barrier is aimed at realizing the rare paths from i to f in order to find out a strange weak value easily. After the symposium, however, the author noticed that a single, free Ising spin system provides a simpler example.

An Ising spin flips up-to-down and down-to-up randomly. A transition probability may be defined by

$$W = \begin{pmatrix} 0 & p \\ p & 0 \end{pmatrix} \text{ and } L = \begin{pmatrix} p & -p \\ -p & p \end{pmatrix}, \quad (32)$$

whose equilibrium state is

$$P_0(x) = \left(\frac{1}{2}, \frac{1}{2}\right) \text{ or } \phi_0(x) = \left(\frac{1}{\sqrt{2}}, \frac{1}{\sqrt{2}}\right).$$

Then, the Hamiltonian for this classical stochastic model is given by [e]

$$H = \begin{pmatrix} p & -p \\ -p & p \end{pmatrix} = p(\sigma_0 - \sigma_x). \quad (33)$$

Another eigenstate, i.e. the excited state is

$$\lambda_1 = 2p \text{ and } \phi_1(x) = \left(\frac{1}{\sqrt{2}}, -\frac{1}{\sqrt{2}}\right),$$

that is, the two eigenstates of H are those of σ_x itself, i.e. the Hadamard states,

$$|0\rangle = \frac{1}{\sqrt{2}}(|\uparrow\rangle + |\downarrow\rangle) \text{ and } |1\rangle = \frac{1}{\sqrt{2}}(|\uparrow\rangle - |\downarrow\rangle). \quad (34)$$

Let us select the initial and the final states as

$$x_i = \uparrow \text{ at } t = 0, \quad x_f = \downarrow \text{ at } t = t_f.$$

Then we find

$$|\psi_i(t)\rangle = |0\rangle + e^{-2pt}|1\rangle, \quad \langle\psi_f(t)| = \langle 0| - e^{-2p(t_f - t)}\langle 1|, \quad (35)$$

and

$$\langle\psi_f|\psi_i\rangle = 1 - e^{-2pt_f}, \quad (36)$$

---

[e] A quantum free spin system has been used often to demonstrate the weak value in quantum mechanics, but it differs from the present system which has a transverse magnetic field p as shown by this Hamiltonian.

by using the eigenvector expansion, where the negative expansion coefficient appears. After the same procedures as the previous model we obtain an extraordinary TTCP,

$$P(0,t) = \frac{\langle \psi_f(t)|0\rangle\langle 0|\psi_i(t)\rangle}{\langle \psi_f|\psi_i\rangle} = \frac{1}{1-e^{-2pt_f}} > 1,$$
$$P(1,t) = \frac{\langle \psi_f(t)|1\rangle\langle 1|\psi_i(t)\rangle}{\langle \psi_f|\psi_i\rangle} = -\frac{e^{-2pt_f}}{1-e^{-2pt_f}} < 0, \tag{37}$$

and a strange weak value,

$$\langle \sigma_x \rangle^{\text{w}}_{(f;i)} = \coth pt_f > 1, \tag{38}$$

again.

On the contrary, when $x_i = x_f = \uparrow$, we find an ordinary TTCP,

$$P(0,t) = \frac{1}{1+e^{-2pt_f}} > 0,$$
$$P(1,t) = \frac{e^{-2pt_f}}{1+e^{-2pt_f}} > 0, \tag{39}$$

and

$$\langle \sigma_x \rangle^{\text{w}}_{(f;i)} = \tanh pt_f < 1.$$

Thus the origin of the negative probability and the strange weak value is more obvious in this simplest system.

5. Extension of TTCP to a density matrix

It should be noted that the physical quantities M_x and A in the previous section are *non-diagonal* in the spin-state representation and have no corresponding quantities in the classical Ising spin system. They are related to the transition rate of the stochastic Ising spin. In order to calculate the expectations of such non-diagonal quantities we need an extension of the TTCP to the two-time conditional density matrix defined by

$$\rho^{\text{w}}_{(f;i)}(t) = \frac{1}{\langle \psi_f|\psi_i\rangle} |\psi_i(t)\rangle\langle \psi_f(t)|$$
$$= \frac{1}{\langle \psi_f|\psi_i\rangle} \sum_{x,x'} \overline{\psi}(x',t|x_f,t_f)\psi(x,t|x_i,0) |x\rangle\langle x'|. \tag{40}$$

From the definition Eq.(18) of the overlap integral $\langle \psi_f|\psi_i\rangle$, it is evident that

$$\text{Tr}\,\rho^{\text{w}}_{(f;i)}(t) = \frac{1}{\langle \psi_f|\psi_i\rangle} \sum_x \overline{\psi}(x,t|x_f,t_f)\psi(x,t|x_i,0) = 1.$$

It should be noted, however, that the diagonal elements of this density matrix are not always positive as is shown by Eq.(28) in Sec.4, when it is diagonalized by using the basis $\{|k\rangle, k = 0, 1, 2, 3\}$, the eigenstates of the Hamiltonian H.

With the use of this density matrix the definition Eq.(22) of the TTCE is extended as

$$\langle Q \rangle^{\mathrm{w}}_{(\mathrm{f};\mathrm{i})} = \mathrm{Tr}\, \rho^{\mathrm{w}}_{(\mathrm{f};\mathrm{i})} Q.$$

Of course this definition of the TTCE results in the classical one, if Q is a diagonal quantity.

The notion of this density matrix has not been used in the conventional classical stochastic process. It should be emphasized, however, that this quantity is within a scheme of the classical stochastic process itself, because the wave functions, ψ and $\overline{\psi}$ in Eq.(40) are related to the forward and the posterior, classical conditional probabilities, respectively. In addition, we have an alternative expression for $\overline{\psi}$,

$$\overline{\psi}(\boldsymbol{x}', t|\boldsymbol{x}_\mathrm{f}, t_\mathrm{f}) = \psi(\boldsymbol{x}', t_\mathrm{f}|\boldsymbol{x}_\mathrm{f}, t)\; \left(= \phi_0(\boldsymbol{x}')^{-1} P(\boldsymbol{x}', t_\mathrm{f} - t|\boldsymbol{x}_\mathrm{f}, 0)\right), \qquad (41)$$

or equivalently,

$$\overline{P}(\boldsymbol{x}', t|\boldsymbol{x}_\mathrm{f}, t_\mathrm{f}) P_0(\boldsymbol{x}_\mathrm{f}) = P(\boldsymbol{x}_\mathrm{f}, t_\mathrm{f}|\boldsymbol{x}', t) P_0(\boldsymbol{x}')$$
$$= P(\boldsymbol{x}', t_\mathrm{f}|\boldsymbol{x}_\mathrm{f}, t) P_0(\boldsymbol{x}_\mathrm{f}), \qquad (42)$$

for $t \leq t_\mathrm{f}$ due to the time-reversal symmetry corresponding to the detailed balance. Then the density matrix Eq.(40) can be written as

$$\rho^{\mathrm{w}}_{(\mathrm{f};\mathrm{i})}(t) = \frac{1}{\langle \psi_\mathrm{f}|\psi_\mathrm{i}\rangle} \sum_{\boldsymbol{x},\boldsymbol{x}'} \frac{P(\boldsymbol{x}', t_\mathrm{f} - t|\boldsymbol{x}_\mathrm{f}, 0) P(\boldsymbol{x}, t|\boldsymbol{x}_\mathrm{i}, 0)}{\phi_0(\boldsymbol{x}')\phi_0(\boldsymbol{x})} |\boldsymbol{x}\rangle\langle \boldsymbol{x}'|, \qquad (43)$$

while the overlap integral can be re-defined by

$$\langle \psi_\mathrm{f}|\psi_\mathrm{i}\rangle = \sum_{\boldsymbol{x}} \frac{P(\boldsymbol{x}, t_\mathrm{f} - t|\boldsymbol{x}_\mathrm{f}, 0) P(\boldsymbol{x}, t|\boldsymbol{x}_\mathrm{i}, 0)}{P_0(\boldsymbol{x})}. \qquad (44)$$

This fact means that we can define the TTCP and the corresponding density matrix with only a pair of the usual, forward conditional probabilities for two individual initial states, $\boldsymbol{x}_\mathrm{i}$ and $\boldsymbol{x}_\mathrm{f}$. We need no data discarding due to the post-selection.

6. Summary and discussions

Except for the facts that the time is imaginary and the wave function is always real and positive, the classical stochastic process can be described in an analogous form of the quantum mechanics, if we use the TTCP. For example, the abnormal behaviors of the weak value in the quantum mechanics are emulated. The TTCP and its TTCE, i.e. the weak values are always real in the present classical case. Therefore, the origin of such abnormal behaviors is clearer than the quantum mechanical case where complex quantities appear.

In addition, if we have not the explicit solution of the eigenvalue problem, we may calculate the weak value at least with use of a Monte-Carlo simulation which is often used to investigate the stochastic model. In performing a simulation it should be noted that we can calculate the TTCP and its TTCE with two usual, forward conditional probabilities for respective *initial* conditions, the pre-selected and the post-selected ones, when the detailed balance condition is satisfied.

The importance of the weak value in the quantum mechanics is that it is related to the new notion of the weak measurement without disturbing the quantum state. An analogous notion of the latter in the classical stochastic process, if any, has not been found yet.

Acknowledgment

This work is supported by Open Research Center Project for Private Universities: Matching fund subsidy from MEXT of Japan.

References

1. Y. Aharonov, D. Z. Albert and L. Vaidman, Phys. Rev. Letters, **60**, 1351, (1988).
2. N. W. M. Ritchie, J. G. Story and R. G. Hulet, Phys. Rev. Letters, **66**, 1107, (1991).
3. A. Hosoya, A lecture note for ORC Summer School on *Decoherence, Entanglement and Entropy* held at Kobe, Japan, 8-11 Aug, 2009.
4. Y. Aharonov, P. G. Bergmann and J.L. Lebowitz, Phys. Rev. **134**, B1410, (1964).
5. A. Hosoya and Y. Shikano, Jour. Phys. A: Math. Theor. **43**, 385307, (2010). (Accurate and detailed references on the present topic are found therein.)
6. R. P. Feynman, 'Negative Probability' in *Quantum Implications*, Routledge & Kegan Paul, London, (1987), p.235-248.
7. D. Sokolovski, Phys. Rev. A **76**, 042125, (2007).
8. Y. Aharonov and A. Botero, Phys. Rev. A, **72**, 052111, (2005).
9. For example, R. Kubo, K. Matsuo and K. Kitahara, Jour. Stat. Phys. **9**, 51, (1973).
10. L. Onsager, Phys. Rev. **37**, 405 (1931).
11. H. Tomita, A. Ito and H. Kidachi, Prog. Theor. Phys. **56**, 786, (1976).

SCALING OF ENTANGLEMENT ENTROPY AND HYPERBOLIC GEOMETRY

HIROAKI MATSUEDA

Advanced Course of Information and Electronic System Engineering,
Sendai National College of Technology,
Sendai, Miyagi 989-3128, Japan
*E-mail: matsueda@sendai-nct.ac.jp

Various scaling relations of the entanglement entropy are reviewed. Based on the scaling, I would like to point out similarity of mathematical formulation among recent topics in wide research area. In particular, the scaling plays crucial roles in identifying a quantum system with a physically different classical system. Close connection between the scaling and hyperbolic geometry and contrast between bulk/edge correspondence and compactification for the identification are also addressed.

Keywords: Entanglement Entropy; Singular Value Decomposition (SVD); Area Law; Black Hole Thermodynamics; Density Matrix Renormalization Group (DMRG); Matrix Product State (MPS); Tensor Product State (TPS) / Projected Entangled Pair State (PEPS); Multiscale Entanglement Renormalization Ansatz (MERA); Hyperbolic Geometry; Anti-de Sitter Space / Conformal Field Theory (AdS/CFT) correspondence; Compactification; Quantum Monte Carlo Simulation (QMC); Image Processing.

1. Introduction

The entanglement entropy is one of the most fundamental concepts in quantum information. Recently, its efficiency as a tool to see through underlying physical principles in our targets is also recognized in statistical physics, condensed matter physics, and string theory. This wide applicability comes from universality of the entropy irrespective of their details. Roughly speaking, the entropy represents a logarithm of a correlation function. Thus, the entropy directly picks up critical exponents. In addition, the entropy can detect topological structure of the manifold where the target model is defined. In order to evaluate the universal feature quantitatively, it is necessary to find scaling relations of the entropy as a function of the linear system size. Furthermore, the scaling relations tell us how physically different systems are associated with each other. When two systems have the same scaling relation, their eigenvalue spectra of the density matrices may be similar in some cases. The identification between the physically different systems leads to deep understanding of duality, holography, and quantum-classical correspondence. They are particularly important ideas in string theory and Quantum Monte Carlo simulation. In ergodic theory, it was a central problem to examine what is a class of measure-preserving transformations in which the information entropy determines isomorphic properties. In viewpoints of symmetry and group theory, the scaling, conformal invariance, and hyperbolic geometry are mutually correlated. Therefore, a lot of important concepts of physics, mathematics, and information merge together in the examination of the entropy. In this article, I would like to review some aspects related to the entropy scaling and its underlying geometrical structure.

2. Entanglement Entropy and Singular Value Decomposition

Let us consider a spatially d-dimensional (dD) quantum system that is composed of a subsystem A and an environment B. The system is sometimes called 'universe' or 'superblock'. The linear size of A is denoted by L. The entanglement entropy represents the amount of information across the boundary between A and B. We start with a pure state of the universe

$$|\psi\rangle = \sum_{x,y} \psi(x,y) |x\rangle |y\rangle, \qquad (1)$$

where $|x\rangle$ and $|y\rangle$ represent basis states of A and B, respectively. The density matrix of the subsystem A is then defined by

$$\rho_A = tr_B |\psi\rangle \langle\psi|, \qquad (2)$$

where the symbol tr_B traces over degrees of freedom inside of B. Then, the entanglement entropy S_A is given by

$$S_A = -tr_A(\rho_A \ln \rho_A). \qquad (3)$$

In order to see physical meaning of the entropy, it is better to introduce the singular value decomposition (SVD) of the wave function $\psi(x,y)$. The SVD of $\psi(x,y)$ is defined by

$$\psi(x,y) = \sum_l U_l(x) \sqrt{\lambda_l} V_l(y), \qquad (4)$$

where $U_l(x)$ and $V_l(y)$ are the column unitary matrices, and λ_l is the singular value that is positively definite. In the following, we use the normalized singular value $p_l = \lambda_l / \sum_l \lambda_l$. Since $\sum_l p_l = 1$, p_l represents a probability of realization of the state labeled by l. The entanglement entropy is then expressed as

$$S = S_A = S_B = -\sum_l p_l \ln p_l. \tag{5}$$

This simple relation $S_A = S_B$ clearly shows non-extensivity of the entanglement entropy, since the volume of A is in general different from that of B. The object that is common in between is their boundary. Then, the entropy would be proportional to the boundary area. This is called area law scaling. In this sense, the entanglement entropy and the thermal entropy behave quite differently. Therefore, to confirm the area law and to find its violation in specific cases are two active topics.

If we assume equal probability condition $p_l = 1/m$ for any l, the entropy yields the standard Boltzmann's law

$$S = -\sum_{l=1}^{m} \frac{1}{m} \ln \frac{1}{m} = \ln m. \tag{6}$$

However, the equal probability condition for the states within the subsystem A is clearly violated in low-dimensional quantum systems. It is easy to see this feature by, for instance, exact diagonalization calculation and other techniques. Thus, the difference between entanglement and thermal entropies is essential. Furthermore, the singular values are uniquely determined after the decomposition, while the column unitary matrices are not. The universal behavior of the entropy is thus due to the presence of a set of the universal singular values.

It is noted that a recent trend to examine the entanglement structure of quantum systems is to introduce the entanglement spectrum as well as the entanglement entropy. The spectrum is defined by

$$E_l = -\ln \lambda_l. \tag{7}$$

Extensive examinations show that the spectrum is a powerful tool to characterize topological nature of the system. The nature is govened by presence or absense of the entanglement gap. The reason for the powerfulness comes from a fact that a topological phase is a phase of matter that cannot be described by an order parameter.

3. Scaling of Entanglement Entropy

3.1. *Historical roots of anomalous entropy scaling*

The thermal entropy is usually an extensive parameter. One exceptional case can be seen in black hole thermodynamics.[1–3] It is well known that the Bekenstein-Hawking entropy is proportional to the surface area $A = 4\pi r^2$ of the event horizon of a black hole:

$$dS = \frac{k_B c^3}{4G\hbar} dA, \qquad (8)$$

Here, $r = 2GM/c^2$ is the Schwartzshild radius. Starting with the surface gravity $\kappa = GM/r^2$, the derivative form of A leads to $dM = (\kappa/8\pi G)dA$. Due to the relativistic energy $E = Mc^2$ and the first law of thermodynamics $dE = TdS$, we obtain $dS = (\kappa c^2/8\pi GT)dA$. If we assume the Hawking temperature $T = \hbar\kappa/2\pi k_B c$, the Eq. (8) is derived. The information absorbed into the black hole can not go out, if we neglect Hawking radiation. Then, a role of the black hole on the entropy is like enviromental degrees of freedom in Eq. (2), since in the definition of the entanglement entropy an observer in A can not access the information in B. Thus, the information theory based on the entanglement entropy has close connection to the black hole physics.

3.2. *General coodinate transformation and horizon*

Consider the metric that does not have any singularities for a static observer. However, an accerated observer may look at different spacetime structure. Actually, the flat Minkowski metric can be transformed into the Rindler one that has event horizon.[4,5] This means that general coodinate transformations beyond Lorentzian ones limits spacetime region that the accerelated observer can access. As shown in Fig. 1, the Rindler observer is confined in the Rindler 'wedge' which acts as a subsystem A, while the outside of the wedge represents the enviromental degrees of freedom B. Thus, the thermal entropy for the field propagating in this geometry behaves as the black hole entropy.

Starting with the Minkowski metric (we consider 1D case for simplicity)

$$ds^2 = -dt^2 + dx^2, \qquad (9)$$

we introduce the following transformation

$$X = \sqrt{x^2 - t^2}, \qquad (10)$$

$$T = \tanh^{-1}\left(\frac{t}{x}\right), \qquad (11)$$

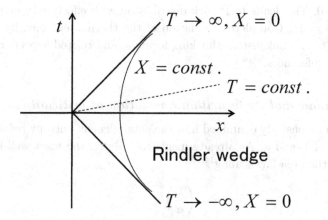

Fig. 1. Minkowski and Rindler coodinates.

or more explicitely

$$t = X \sinh T, \tag{12}$$
$$x = X \cosh T, \tag{13}$$

where $-\infty < T < \infty$, and $0 < X < \infty$. Then, the metric for the new coodinate is given by

$$ds^2 = -X^2 dT^2 + dX^2. \tag{14}$$

We clearly see that the accesible region is only the right Rindler wedge as shown in Fig. 1.

As a tutorial example, let us consider the massless scalar field ϕ propagating in the Rindler space. The action is given by

$$I = \int dT dX \sqrt{-g} \left(-\frac{1}{2} g^{\mu\nu} \partial_\mu \phi \partial_\nu \phi \right). \tag{15}$$

The Lagrange equation is obtained as

$$\frac{\partial^2}{\partial T^2} \phi = \left(X^2 \frac{\partial^2}{\partial X^2} + X \frac{\partial}{\partial X} \right) \phi, \tag{16}$$

and yields the following solution

$$\phi(T, X) = A e^{i\Omega(t-x)} = A \exp\left(i\Omega X e^{-T} \right). \tag{17}$$

By taking the inverse Fourier transform with respect to T, we see that the power spectrum of this field is not monochromatic

$$\int_{-\infty}^{\infty} dT \exp\left(i\Omega X e^{-T} \right) e^{i\omega T} = (-i\Omega X)^{i\omega} \Gamma(-i\omega), \tag{18}$$

where $\omega \neq 0$. This leads to Planck distribution with effective inverse temperature $\beta = 2\pi$. Complete calculation of the thermal entropy, its correspondence to the Bekenstein-Hawking formula, and related aspects can be seen in the references.[6–10]

3.3. *Area law and its logarithmic violation at criticality*

It has been extensively examined how the entanglement entropy behaves as a function of L and d. As already mentioned above, the most well-known formula is the area-law scaling[11,12]

$$S \propto \left(\frac{L}{a}\right)^{d-1}, \tag{19}$$

which tells us non-extensivity of S in contrast to the thermal entropy. This relation is strictly satisfied in gapped quantum systems.

The violation of the area law occurs in cases of 1D critical systems.[13–16] The entropy is given by

$$S = \frac{1}{3} c \ln \frac{L}{a}, \tag{20}$$

where c is the central charge and a is lattice cutoff. This was first derived from the conformal field theory (CFT), and then numerically confirmed in various models of statistical physics.[17–19] Away from a critical point, the entropy is deformed as

$$S = \frac{1}{6} c \mathcal{A} \ln \frac{\xi}{a}, \tag{21}$$

with correlation length ξ and the number of boundary points $\mathcal{A}$ of A.

3.4. *Fermi surfaces and entropy*

The violation also occurs in models with the Fermi surfaces.[20–24] The Fermi surface devides the momentum space into two sectors: sets of occupied and unoccupied states. The electronic excitations normal to the Fermi surface are like those of chiral Luttinger liquid. The excitations lead to logarithmic correction. The entropy is then given by

$$S = \frac{1}{3} C \left(\frac{L}{a}\right)^{d-1} \ln \frac{L}{a} + B \left(\frac{L}{a}\right)^{d-1} + A \left(\frac{L}{a}\right)^{d-2} + \cdots, \tag{22}$$

$$C = \frac{1}{4(2\pi)^{d-1}} \int_{\partial\Omega} \int_{\partial\Gamma} |n_x \cdot n_p| \, dA_x dA_p, \tag{23}$$

where C is the number of excitation modes across the Fermi surface $\partial\Gamma$, $\partial\Omega$ is the spatial region considered, and n_x and n_p are the unit normals to the boundaries.

3.5. *Topological entanglement entropy*

The physical characterization of topological orders of 2D is a hot topic in condensed matter physics. Usually, the topological nature can be seen as edge excitations and quasiparticle statistics. However, the topological order is manifest in the basic entanglement of the ground state wave function.[25,26] In Z_q lattice gauge theory, the entropy has a subleading term called topological entropy in addition to the area-law scaling

$$S = \alpha L - \ln\sqrt{q} + \cdots . \qquad (24)$$

The string-net condensed model and the Kitaev toric code are two important examples of realizing the presence of the topological entropy.

3.6. *Entanglement support of matrix product state*

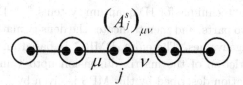

Fig. 2. MPS on 1D chain. Each entangled bond is characterized by a symbol $\mu = 1, 2, ..., \chi$. The index s is local degree of freedom ($s = \uparrow, \downarrow$ in $S = 1/2$ spin systems).

In statistical physics, there is another type of entropy scaling originated from a recently developed variational approach to quantum systems. The approach optimizes the following matrix product state (MPS) ansatz[27,28]

$$|\psi\rangle = \sum_{\{s_j\}} tr(A_1^{s_1} A_2^{s_2} \cdots A_n^{s_n}) |s_1 s_2 \cdots s_n\rangle, \qquad (25)$$

where the site-dependent matrices $A_j^{s_j}$ have $\chi \times \chi$ dimensions, and s_j represents local degrees of freedom (see Fig. 2). Historically, matrix product forms play crucial roles in exacly solvable statistical models.

Let us consider the simplest case. We are going to express the singlet state $|\psi\rangle = |\uparrow\downarrow\rangle - |\downarrow\uparrow\rangle$ by a particular MPS. It is clear that the direct product state (local decomposition) is not exact

$$|\phi\rangle = \sum_{s_1=\uparrow,\downarrow} c_1^{s_1}|s_1\rangle \otimes \sum_{s_2=\uparrow,\downarrow} c_2^{s_2}|s_2\rangle \qquad (26)$$

$$= c_1^\uparrow c_2^\uparrow |\uparrow\uparrow\rangle + c_1^\uparrow c_2^\downarrow |\uparrow\downarrow\rangle + c_1^\downarrow c_2^\uparrow |\downarrow\uparrow\rangle + c_1^\downarrow c_2^\downarrow |\downarrow\downarrow\rangle. \qquad (27)$$

Actually, we can not simultaneously take $c_1^\uparrow c_2^\uparrow = 1$, $c_1^\uparrow c_2^\downarrow = 0$, $c_1^\downarrow c_2^\uparrow = 0$, and $c_1^\downarrow c_2^\downarrow = -1$. However, we can introduce the following expression

$$|\psi\rangle = \sum_{s_1,s_2} A^{s_1} B^{s_2} |s_1 s_2\rangle \qquad (28)$$

where the two local vectors A^{s_1} and B^{s_2} are taken to be

$$A^\uparrow = (x\ y),\ A^\downarrow = (z\ w), \qquad (29)$$

$$B^\uparrow = \frac{1}{xw-yz}\begin{pmatrix} y \\ -x \end{pmatrix}, B^\downarrow = \frac{1}{xw-yz}\begin{pmatrix} w \\ -z \end{pmatrix}, \qquad (30)$$

for scalar variables x, y, z, and w. This expression means that the matrix dimension χ provides quantum correlation between neighboring spins.

The MPS state is known to be exact for gapped 1D systems such as the valence bond solid (VBS) state. Furthermore, the MPS is foundation of the density matrix renormalization group (DMRG) method that is a powerful numerical technique for 1D quantum systems.[29,30] DMRG devides the system into two parts, and then truncates the density matrix eigenstates with the small eigenvalues. Therefore, DMRG is based on entanglement control. The restriction of the matrix dimension upto χ means that the amount of information described by this MPS is given by

$$S_\chi = -\sum_{l=1}^\chi \lambda_l \ln \lambda_l, \qquad (31)$$

where $\lambda_1 > \lambda_2 > \cdots > \lambda_\chi$. If we assue that $A_j^{s_j}$ is a scalar variable ($\chi = 1$), Eq (25) gives a local approximation as I have already mentioned. Taking a sufficiently large χ value gives us an asymptotically exact wave function. Thus, the matrix dimension χ represents how precisely we can take long-range quantum correlation. When we apply this variational method to 1D critical systems, the half-chain entropy ($\mathcal{A} = 1$) behaves as[31–33]

$$S_\chi = \frac{c\kappa}{6}\ln\chi, \qquad (32)$$

where the exponent κ is defined by

$$\kappa = \frac{6}{c\left(\sqrt{12/c}+1\right)}. \qquad (33)$$

Comparing Eq. (32) with Eq. (21), we know that

$$\xi = \chi^\kappa, \qquad (34)$$

which means that χ is actually related to the length scale ξ. For instance, in the Ising universality class $c = 1/2$

$$\kappa \sim 2, \qquad (35)$$
$$S_\chi \sim \frac{1}{6} \ln \chi. \qquad (36)$$

I have found an old DMRG paper that addresses the correlation length ξ as a function of the truncation number.[34] The result for 1D free fermion with the central charge $c = 1$ is given by

$$\xi = -\frac{1}{\ln|1 - k\chi^{-\beta}|} = \frac{1}{k}\chi^\beta, \qquad (37)$$

with $\beta \sim 1.3$ and $k \sim 0.45$. This is really consistent with the above argument where $\kappa \sim 1.33$ for $c = 1$.

4. Holographic Entanglement Entropy: Connection between Scaling and Hyperbolic Geometry

4.1. *Anti-de Sitter space*

In superstring theory, it has been an interesting topic to find the fundamental mechanism of the anti-de Sitter space (AdS) / CFT correspondence. This is holographic correspondence between a quantum theory, CFT_{d+1}, and general relativity on the AdS space, AdS_{d+2} (Here, CFT_{d+1} is abbreviation of spatially d-dimensional conformal field theory with one time axis). The AdS space is $(d+2)$-dimensional hyperbolic surface embedded in one higher dimensional flat space $E^{d+1,2}$

$$-X_{-2}^2 - X_{-1}^2 + \sum_{k=0}^{d} X_k^2 = -l^2, \qquad (38)$$

(see Fig. 3). This space has two time-like directions X_{-2} and X_{-1}. The metric of the AdS space can be represented as

$$ds^2 = g_{\mu\nu} dx^\mu dx^\nu \qquad (39)$$
$$= \frac{l^2}{z^2} \left(dz^2 + \eta_{ij} dx^i dx^j \right), \qquad (40)$$

where l is the curvature of this space. The index z is called radial axis, and i runs over $0, 1, 2, ..., d$. The AdS space is a vacuum solution of the Einstein equation

$$G^{\mu\nu} - \frac{1}{2}g^{\mu\nu}R + g^{\mu\nu}\Lambda = \kappa T^{\mu\nu} = 0, \qquad (41)$$

with negative cosmological constant $\Lambda = -d(d+1)/2l^2$.

4.2. *Killing equation and conformal invariance*

The Killing equation on the bulk AdS space approaches the conformal one at the boundary $z \to 0$. Let us start from infinitesimal transformation

$$x'^{\mu} = x^{\mu} + \xi^{\mu}(x), \qquad (42)$$

and then the new line element is given by

$$ds'^2 = ds^2 + (\nabla_\mu \xi_\nu + \nabla_\nu \xi_\mu)\, dx^\mu dx^\nu, \qquad (43)$$

where the covariant derivative is defined by

$$\nabla_\mu \xi_\nu = \partial_\mu \xi_\nu - \Gamma^\lambda_{\mu\nu}\xi_\lambda, \qquad (44)$$

with the Christoffel symbol

$$\Gamma^\rho_{\mu\nu} = \frac{1}{2}g^{\rho\tau}\left(\partial_\mu g_{\nu\tau} + \partial_\nu g_{\mu\tau} - \partial_\tau g_{\mu\nu}\right). \qquad (45)$$

Here, the isometory $ds'^2 = ds^2$ yields the Killing equation $\nabla_\mu \xi_\nu + \nabla_\nu \xi_\mu = 0$. We consider the following transformation

$$\xi^i(z, x) \to \epsilon^i(x), \qquad (46)$$
$$\xi^z(z, x) \to z\zeta(x), \qquad (47)$$

where the boundary does not move after the transformation. Substituting these equations and the AdS metric into the Killing equation, we actually obtain the conformal Killing equation at the boundary

$$\partial_i \epsilon_j + \partial_j \epsilon_i = \zeta \eta_{ij}. \qquad (48)$$

Thus, the CFT lives on the boundary of the AdS space. Actually, it is possible for various models to prove

$$\left.\frac{\delta}{\delta\phi(x)}\frac{\delta}{\delta\phi(x')}\exp\left(-\frac{1}{2\kappa^2}I\right)\right|_{\phi=0} \propto \frac{1}{|x-x'|^\Delta}, \qquad (49)$$

with the scaling dimension Δ of the CFT and I denoting the classical Einstein-Hilbert action. This is so called Gubser-Krevanov-Polyakov-Witten relation.[35–38]

4.3. Geometric interpretation of entanglement entropy

An important guiding result to understand the physical background of the AdS/CFT correspondence is the so called Ryu-Takayanagi formula for the entanglement entropy[39-41]

$$S = \frac{\text{Area}(\gamma_A)}{4G}, \qquad (50)$$

where γ_A is the minimal surface whose boundary is given by the manifold $\partial \gamma_A = \partial A$, and G is the Newton constant of gravity in the AdS space. This is general extension of the Bekenstein-Hawking entropy in Eq. (8).

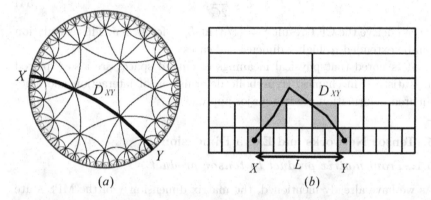

Fig. 3. Two representations of 2D hyperbolic space and geodesic line: (a) Poincaré disk model, (b) discrete model. In (b), D_{XY} may not look like a geodesic line, but this is due to ambiguity of lattice discretization. The original AdS metric in Eq. (40) can be transformed into $ds^2 = (d\tau \log 2)^2 + (2^{-\tau/l} dx)^2$. This form is comparable to Fig. (b).

For $d = 1$, Area(γ_A) is the geodesic distance D_{XY} between $X = (-L/2, a)$ and $Y = (L/2, a)$ for $a \to 0$ as shown in Fig. 3. These points are located on the boundary of the AdS space. Let us derive an explicit formula of D_{XY}, since this quantity plays a crucial role in the later discussion. The geodesic equation is given by

$$\frac{d^2 x^\rho}{d\tau^2} + \Gamma^\rho{}_{\mu\nu} \frac{dx^\mu}{d\tau} \frac{dx^\nu}{d\tau} = 0, \qquad (51)$$

and the solution is the half-cycle

$$(x^1, x^2) = (x, z) = \frac{L}{2}(\cos\theta, \sin\theta), \qquad (52)$$

where $d\theta/dt = \sin\theta$, $\epsilon \leq \theta \leq \pi - \epsilon$, and $(L/2)\sin\epsilon = a$. By using the equation for the geodesic line, we calculate D_{XY} as

$$\begin{aligned}D_{XY} &= 2\int_\epsilon^{\pi/2} \frac{l}{z} d\theta \sqrt{(\partial_\theta z)^2 + (\partial_\theta x)^2} \\ &= l \ln\left(\frac{1+\cos\epsilon}{1-\cos\epsilon}\right) \\ &= 2l \ln\left(\frac{L + \sqrt{L^2 - (2a)^2}}{2a}\right).\end{aligned} \quad (53)$$

By combining this with the Brown-Henneaux central charge[42]

$$c = \frac{3l}{2G}, \quad (54)$$

we can derive the CFT result $S = (c/3)\ln(L/a)$ for $L \gg a$. This calculation can be extended to higher dimensional cases.

It is noted that physical meanings of curved spaces are also explored in statistical mechanics. Hyperbolic deformation of interaction and sine-squared deformation are two major approaches.[43–46]

5. Tensor Networks and Extra Dimension

5.1. *From matrix product to tensor product*

As we have already mentioned, the matrix dimension χ of the MPS state represents how strong quantum entanglement appears. However, the MPS state is appropriate only in 1D gapped cases. Direct extention of this approach to higher dimensions is to construct tensor networks (or sometimes called tensor product state, TPS, and projected entangled-pair state, PEPS).[47–52] We show a schematic viewgraph of TPS in Fig. 4. We define a tensor, $T^{s_j}_{\mu\nu\lambda\ldots}$, on each site j. The number of the indices, $\mu, \nu, \lambda, \cdots$, of each tensor corresponds to that of connected bonds. If each bond is maximally entangled, the state of the bond can be represented by α and its ancila $\bar{\alpha}$

$$|\phi\rangle = \sum_{\alpha=1}^{\chi} \frac{1}{\sqrt{\chi}} |\alpha\bar{\alpha}\rangle. \quad (55)$$

Then, the entropy for each bond is given by $S_{bond} = -tr(\rho\ln\rho) = \log\chi$, where the trace is taken over the ancila degree of freedom. The entanglement entropy of the subsystem marked by a dotted line in Fig. 4 is given by

$$S = NS_{bond} \sim \left(\frac{L}{a}\right)^{d-1} \ln\chi, \quad (56)$$

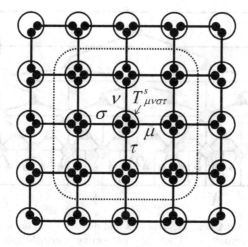

Fig. 4. Tensor network structure on 2D square lattice. The tensor $T^s_{\mu\nu\sigma\tau}$ with linear dimension χ is defined on each lattice site. The quantum entanglement between neighboring sites is represented by the indices μ, ν, σ, and τ.

where N is the number of the entangled bonds, L is linear size, and a is lattice constant. This relation is nothing but the area law scaling. Thus, we know that the TPS state is suitable for d-dimensional gapped cases. On ther other hand, $\chi \sim L$ in critical cases, and further improvement of the network structure may be required. This is the purpose of the next section where tree tensor and MERA networks are introduced.

5.2. *Tree tensor networks and multiscale entanglement renormalization ansatz: Hierarchical tensor network*

Constructing efficient variational optimization methods of MPS and TPS is of practical importance in condenced matter physics. An advantage of these methods is to keep the area-law scaling of the entropy, and thus is suitable to gapped d-dimensional systems. The application of tensor networks to critical systems leads to the tree tensor network (TTN) and the multiscale entanglement renormalization ansatz (MERA).[53–60] They are spatially $(d+1)$-dimensional hierarchical networks that are also compatible with real-space renormalization group. In particular, the MERA network forms discrete AdS space as shown in Figs. 5 and 6.[61,62] Although strict correspondence between AdS/CFT duality and MERA is not constructed in the present stage, they are very similar. When we express a critical system by using TPS, we need large tensor dimension $\chi \sim L$. On the other

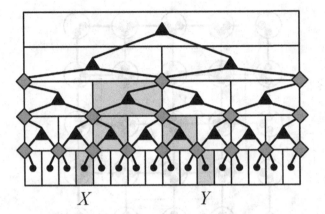

Fig. 5. Schematic MERA network. Triangles and diamonds are isometory tensors and disentangler tensors, respectively. Each bond connecting a triangle with a diamond represents tensor product. For comparison, Fig. 3 is combined with the network.

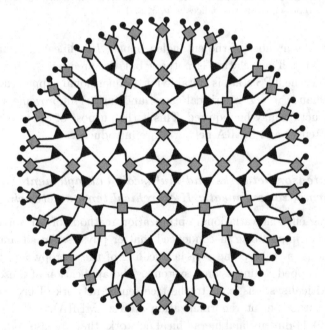

Fig. 6. Deformation of MERA network in Fig. 5.

hand, each tensor in the hierarchical network of MERA has relatively small dimension. Thus, we may say that MERA constructs more classical-like states in the holographic space.

6. Compactification and Entropy

6.1. *Bulk/edge correspondence and compactification*

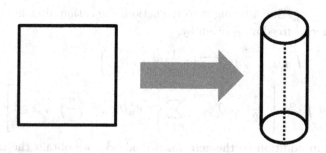

Fig. 7. Compactification of open 2D sheet into 1D system.

AdS/CFT is a kind of bulk/edge correspondence on non-compact manifold. Compactification of non-compact space is an alternative way to find correspondence between physically different systems. The comparison between these ways may be useful for further examination of the entanglement entropy. The concept of the compactification is schematically shown in Fig. 7 where an open 2D sheet is rolled up. When our spatial resolution is worse than the compactification radius, this is effectively a 1D system. Unfortunately, we can not directly define the entanglement entropy in the classical space, since entanglement represents quantum correlation. However, according to the Ryu-Takayanagi's formula, a geometric quantity would correspond to the entropy. Then, the problem is to find geometrical meaning of the entanglement entropy in the non-compact manifold before compactification. In the following subsections, some historical review and my recent work closely related to this aspect are presented.

6.2. *Field theory with extra dimension*

Let us look at compactification of a massless scalar field model. The metric is given by the flat 4D Minkowski one ($i = 0, 1, 2, 3$) plus additional component $dx^4 = ad\theta$

$$g_{\mu\nu} = \begin{pmatrix} \eta_{ij} & 0 \\ 0 & a^2 \end{pmatrix}, \tag{57}$$

where the radial direction is represented by θ, and a is its radius. We assume a cylindrical boundary condition for the θ direction, and the others are not.

Then, we have $0 \leq \theta \leq 2\pi$. The scalar field $\phi(x, \theta)$ is discretized as

$$\phi(x, \theta) = \frac{1}{\sqrt{2\pi}} \sum_n \phi_n(x) e^{in\theta}, \tag{58}$$

where $\phi_{-n} = \phi_n^*$. Substituting it to the action, the action after integrating over θ degree of freedom is given by

$$I = \frac{1}{2\pi} \int d^d x \int_0^{2\pi} d\theta \sqrt{-g} \left(-\frac{1}{2} g^{\mu\nu} \partial_\mu \phi \partial_\nu \phi \right) \tag{59}$$

$$= a \int d^d x \left[-\frac{1}{2} \partial_i \phi_0 \partial^i \phi_0 - \sum_{n \geq 1} \left\{ \partial_i \phi_n^* \partial^i \phi_n + \left(\frac{n}{a}\right)^2 \phi_n^* \phi_n \right\} \right]. \tag{60}$$

Therefore, in addition to the zero mass mode ϕ_0, we obtain the massive modes ϕ_n with mass $M_n = n/a$. They are called Kaluza-Klein modes. In the classical field theory which aims dimensional reduction, a should be taken to be small enough. However, in the following, we are interested in cases of various a values. For large a values, these modes tend to be gapless. Thus, this indicates that the enough internal degree of freedom is necessary to describe low-energy excitations. This feature can also be seen in the tensor network formulation of quantum states.

6.3. VBS/CFT correspondence

Recently, it has been shown that the entanglement spectra of the VBS state on 2D lattices are closely related to a thermal density matrix of a holographic spin chain whose spectrum is reminiscent of that of the spin-1/2 Heisenberg chain.[63,64] Here, we briefly touch on an idea of this correspondence based on the tensor product. It might be possible to have a different view based on open/close string duality. We start with the 2D VBS state on $M \times N$ lattice (vertical × horizontal) expressed by

$$|\psi\rangle = \sum_I c_I |I_1 I_2 \cdots I_N\rangle, \tag{61}$$

where $I_n = (i_{1,n}, i_{2,n}, ..., i_{M,n})$. Let us consider a cylindrical boundary condition that the vertical direction is rolled up, while the horizontal axis remains open. Then, we can introduce the coefficient c_I defined by

$$c_I = \sum_\Lambda L^{I_1}_{\Lambda_1} B^{I_2}_{\Lambda_1 \Lambda_2} \cdots B^{I_{N-1}}_{\Lambda_{N-2} \Lambda_{N-1}} R^{I_N}_{\Lambda_{N-1}}, \tag{62}$$

where $\Lambda_n = (\alpha_{1,n}, \alpha_{2,n}, ..., \alpha_{M,n})$ and $\alpha_{j,n} = 1, 2, ..., \chi$ with bond dimension χ of the original VBS. The periodic boundary condition along the vertical direction due to the cylinder form is thus expressed by

$$B^{I_n}_{\Lambda_{n-1}\Lambda_n} = tr \prod_{j=1}^{M} A^{i_{j,n}}_{\alpha_{j,n-1}\alpha_{j,n}}, \tag{63}$$

and its boundary terms $L^{I_1}_{\Lambda_1}$ and $R^{I_N}_{\Lambda_{N-1}}$. Clearly, Eq. (62) is a MPS form. When we devide this system into two parts, the state $|\psi\rangle$ can be expressed by the Schmidt decomposition

$$|\psi\rangle = \sum_{I_a,I_b} \sum_\Lambda L^{I_a}_\Lambda R^{I_b}_\Lambda |I_a, I_b\rangle, \tag{64}$$

where $L^{I_a} = L^{I_1} B^{I_2} \cdots B^{I_l}$ and $R^{I_b} = B^{I_{l+1}} \cdots B^{I_{N-1}} R^{I_N}$. The dimension of Λ is χ^M. Originally, the VBS state has relatively small dimension χ due to the presence of the Haldane gap. However, χ^M becomes very large, leading to critical behavior. I think it's interesting to compare this idea with the compactification discussed in the previous subsection.

6.4. Suzuki-Trotter decomposition

The readers expert for quantum Monte Carlo (QMC) simulation may be aware of the similarity between Suzuki-Trotter decomposition (STD) and compactification.[65] The correspondence between classical and quantum systems are induced by the following STD

$$e^{X+Y} = \lim_{M\to\infty} \left(e^{\frac{X}{M}} e^{\frac{Y}{M}}\right)^M \tag{65}$$

for non-commutative operators X and Y. In some cases, the right hand side can be treated exactly.

It is well-known that the partition function of the transverse-field Ising chain can be mapped onto that of the 2D classical anisotropic Ising model. The transverse-field Ising Hamiltonian is given by

$$H = H_0 + H' = -J \sum_{i=1}^{L} \sigma^z_i \sigma^z_{i+1} - \lambda \sum_{i=1}^{L} \sigma^x_i. \tag{66}$$

Let us derive the classical model with use of the STD. We introduce the partition function $Z = \text{Tr}e^{-\beta H}$, and decompose it into M slices as

$$Z = \sum_{\{\sigma_1\}} \langle\{\sigma_1\}| \left[\exp\left(-\frac{\beta}{M}H_0\right)\exp\left(-\frac{\beta}{M}H'\right)\right]^M |\{\sigma_1\}\rangle \qquad (67)$$

$$= \sum_{\{\sigma_1\},...,\{\sigma_M\}} \prod_{k=1}^{M} \langle\{\sigma_k\}|\exp\left(-\frac{\beta}{M}H_0\right)\exp\left(-\frac{\beta}{M}H'\right)|\{\sigma_{k+1}\}\rangle \qquad (68)$$

$$= \sum_{\{\sigma_1\},...,\{\sigma_M\}} \prod_{k=1}^{M} \exp\left\{\frac{\beta}{M}J\sum_{i=1}^{L}\sigma_i^k\sigma_{i+1}^k\right\}$$

$$\times \langle\{\sigma_k\}|\exp\left(-\frac{\beta}{M}H'\right)|\{\sigma_{k+1}\}\rangle. \qquad (69)$$

Here, we have imposed the cylindrical boundary condition, $\{\sigma_1\} = \{\sigma_{M+1}\}$. Thus, we may say that the mathematical processes of the STD zoom up internal degrees of freedom on each site, which is analogous to the previous sections. For $\sigma, \sigma' = \pm 1$, we have a relation

$$\langle\sigma|e^{\frac{\beta}{M}\lambda\sigma^x}|\sigma'\rangle = A\exp\left\{-\frac{1}{2}\sigma\sigma'\ln\left(\tanh\frac{\beta}{M}\lambda\right)\right\}, \qquad (70)$$

with $A = \sqrt{(1/2)\sinh(2\beta/M)\lambda}$. Thus, Z is expressed as

$$Z = A^M \prod_{k=1}^{M} \exp\left\{\frac{\beta}{M}J\sum_{i=1}^{L}\sigma_i^k\sigma_{i+1}^k\right\}$$

$$\times \exp\left\{-\frac{1}{2}\sum_{i=1}^{L}\sigma_i^k\sigma_i^{k+1}\ln\left(\tanh\frac{\beta}{M}\lambda\right)\right\}, \qquad (71)$$

and finally the effective Hamiltonian is given by

$$H_{\text{eff}} = \sum_{i=1}^{L}\sum_{k=1}^{M}\left(J_1\sigma_i^k\sigma_{i+1}^k + J_2\sigma_i^k\sigma_i^{k+1}\right), \qquad (72)$$

where $J_1 = -J/M$ and $J_2 = -(1/2\beta)\ln(\tanh(\beta\lambda/M))$.

6.5. *Entropy scaling, quantum-classical correspondence, and hyperbolic geometry hidden in image processing based on SVD*

According to the previous discussion, let us finally examine what is a geometrical object corresponding to the entropy on the classical spin systems before compactification. That is a snapshot of a particular spin configuration at criticality.

We regard the orignal image data of the snapshot with $M \times N$ pixels as a matrix $\psi(x,y)$ ($1 \le x \le M$, $1 \le y \le N$). We assume that ψ is real. Then, it is possible to introduce the density matrices

$$\rho_X(x,x') = \sum_y \psi(x,y)\psi(x',y), \tag{73}$$

$$\rho_Y(y,y') = \sum_x \psi(x,y)\psi(x,y'). \tag{74}$$

Their L non-zero eigenvalues are the same ($L = \min(M,N)$). According to Eqs. (4) and (5), we can define the entropy. In order to intuitively understand scaling relations which this snapshot entropy satisfies, let us imagine a snapshot of the classical 2D isotropic Ising spin system at criticality. There, the spin configuration is fractal, since various sizes of the ordered clusters coexist due to the critical fructuation. This also means that various length scales are mixed. We can pick up many fractal subsystems from the original snapshot, but their patterns themselves are different with each other. Therefore, after correcting all of the possible patterns, they would cover the total information of the partition function. The information of thermal fructuation contained in the partition function is represented by the differen patterns. In that sense, only one snapshot is necessary at criticality. According to the STD, we expect that there exists a quantum 1D system that is transformed into 2D isotropic classical Ising model. Therefore, the snapshot entropy should obey the scaling relation equal to that of the critical 1D quantum systems. Actually, I have recently shown that this conjecture is at least correct in the 2D isotropic classical Ising model on $L \times L$ lattice.[66] I have obtained the following results at T_c

$$S = -\sum_{l=1}^{L} \lambda_l \ln \lambda_l = \ln L - 2, \tag{75}$$

$$S_\chi = -\sum_{l=1}^{\chi<L} \lambda_l \ln \lambda_l \sim \frac{1}{6}\ln \chi + \gamma, \tag{76}$$

for sufficiently large L and a positive γ value. They are comparable to CFT and MPS results. Furthermore, when we visualize each layer of the SVD

$$\psi^{(l)}(x,y) = U_l(x)\sqrt{\Lambda_l}V_l(y), \tag{77}$$

we know that $\psi^{(l)}(x,y)$ has its own length scale. The images of $\psi^{(l)}(x,y)$ with larger singular values contain more global spin structures. Extensive examinations suggest that the index l corresponds to the radial direction of the discrete AdS space. The result means that the coarse graining of the image by tuncating small singular values correspondes to a flow from the boundary to the bulk on the AdS space. This would be related to the holographic renormalization group.

The entropy of more realistic images also obey a clear scaling relation analogous to the entanglement support of MPS.[67] This is surprizing, since we think it's impossible to determine 'the central charge' of the realistic images. We need more detailed analysis of this type of quantum-classical correspondence.

7. Summary

In this article, I have reviewed various scaling relations of the entanglement entropy and related topics. Since the entropy represents universality of the model considered, we can obtain more global viewpoints beyond the model itself. In particular, the scaling is a very powerful tool to look at quantum-classical correspondence. Finally, we should be carefull to a fact that the quality of the information is given by the wave function, not the entropy. In this respect, we need comprehensive study of the scaling analysis and numerical techniques of optimizing tensor networks.

References

1. Jacob D. Bekenstein, *"Black Holes and Entropy"*, Phys. Rev. D **7**, 2333 (1973).
2. S. W. Hawking, *"Particle Creation by Black Holes"*, Commun. Math. Phys. **43**, 199 (1975).
3. S. W. Hawking, *"Black holes and thermodynamics"*, Phys. Rev. D **13**, 191 (1976).
4. Wolfgang Rindler, *"Relativity: Special, General, and Cosmological"*, OXFORD.
5. P. C. W. Davies, *"Scalar particle production in Schwartzschild and Rindler metrics"*, J. Phys. A: Math. Gen. **8**, 609 (1975).
6. Leonard Susskind and John Uglum, *"Black Hole Entropy in Canonical Quantum Gravity and Superstring Theory"*, Phys. Rev. D **50**, 2700 (1994).
7. D. Kobat and M. J. Strassler, *"A Comment on Entropy and Area"*, Phys. Lett. B **329**, 46 (1994).
8. R. Emparan, *"Heat kernels and thermodynamics in Rindler space"*, Phys. Rev. D **51**, 5716 (1995).
9. T. Padmanabhan, *"Thermodynamical Aspects of Gravity: New insights"*, rept. Prog. Phys. **73**, 046901 (2010).
10. Ted Jacobson, *"Thermodynamics of Spacetime: The Einstein Equation of State"*, Phys. rev. Lett. **75**, 1260 (1995).
11. Luca Bombelli, Rabinder K. Koul, Joohan Lee, and Rafael D. Sorkin, *"Quantum source of entropy for black holes"*, Phys. Rev. D **34**, 373 (1986).

12. M. Srednicki, "*Entropy and area*", Phys. Rev. Lett. **71**, 666 (1993).
13. Christoph Holzhey, Finn Larsen, Frank Wilczek, "*Geometric and renormalized entropy in conformal field theory*", Nucl. Phys. B **424**, 443 (1994).
14. P. Calabrese and J. Cardy, "*Entanglement Entropy and Quantum Field Theory*", J. Stat. Mech. **0406**, P002 (2004) [note added: arXiv:hep-th/0405152].
15. Pasquale Calabrese and John Cardy, "*Entanglement entropy and conformal field theory*", J. Phys. A **42**, 504005 (2009).
16. J. Jacobsen, S. Ouvry, V. Pasquier, D. Serban, and L. F. Cugliandolo eds. "Les Houches 2008, Session LXXXIX, Exact Methods in Low-Dimensional Statistical Physics and Quantum Computing", OXFORD (2009).
17. M. B. Plenio, J. Eisert, J. Dreißig, and M. Cramer, "*Entropy, Entanglement, and Area: Analytical Results for harmonic Lattice Systems*", Phys. Rev. Lett. **94**, 060503 (2005).
18. A. Riera and J. I. Latorre, "*Area law and vacuum reordering in harmonic networks*", Phys. Rev. A **74**, 052326 (2006).
19. G. Vidal, J. I. Latorre, E. Rico, and A. Kitaev, "*Entanglement in Quantum Critical Phenomena*", Phys. Rev. Lett. **90**, 227902 (2003).
20. Michael M. Wolf, "*Violation of the Entropic Area Law for Fermions*", Phys. Rev. Lett. **96**, 010404 (2006).
21. Dimitri Gioev and Israel Klich, "*Entanglement Entropy of Fermions in Any Dimension and the Widom Conjecture*", Phys. Rev. Lett. **96**, 100503 (2006).
22. Dimitri Gioev, "*Szegö limit theorem for operators with discontinuous symbols and applications to entanglement entropy*", arXiv:math/0212215.
23. Weifei Li, Letian Ding, Rong Yu, Tommaso Roscilde, and Stephan Haas, "*Scaling behavior of entanglement in two- and three-dimensional free-fermion systems*", Phys. Rev. B **74**, 073103 (2006).
24. T. Barthel, M-.-C. Chung, and U. Schollwöck, "*Entanglement scaling in critical two-dimensional femrionic and bosonic systems*", Phys. Rev. A **74**, 022329 (2006).
25. Alexei Kitaev and John Preskill, "*Topological Entanglement Entropy*", Phys. Rev. Lett. **96**, 110404 (2006).
26. Michael Levin and Xiao-Gang Wen, "*Detecting Topological Order in a Ground State Wave Function*", Phys. Rev. Lett. **96**, 110405 (2006).
27. S. Östlund and S. Rommer, "*Thermodynamics Limit of Density Matrix Renormalization*", Phys. Rev. Lett. **75**, 3537 (1995).
28. S. Rommer and S. Östlund, "*class of ansatz wave functions for one-dimensional spin systems and their relation to the density matrix renormalization group*", Phys. Rev. **B55**, 2164 (1997).
29. S. R. White, "*Density matrix formaulation for quantum renormalization group*", Phys. Rev. Lett. **69**, 2863 (1992).
30. S. R. White, "*Density-matrix algorithms for quantum renormalization group*", Phys. Rev. B **48**, 10345 (1993).
31. L. Tagliacozzo, Thiago. R. de Oliveira, S. Iblisdir, and J. I. Latorre, "*Scaling of entanglement support for matrix product states*", Phys. Rev. B **78**, 024410 (2008).
32. Frank Pollman, Subroto Mukerjee, Ari M. Turner, and Joel E. Moore, "*The-

ory of Finite-Entanglement Scaling at One-Dimensional Quantum Critical Points", Phys. Rev. Lett. **102**, 255701 (2009).
33. Ching-Yu Huang and Feng-Li Lin, "*Multiparticle entanglement measures and quasntum criticality from matrix and tensor product states*", Phys. Rev. A **81**, 032304 (2010).
34. Martin Andersson, Magnus Boman, and Stellan Östlund, "*Density-matrix renormalization group for a gappless system of free fermions*", Phys. Rev. B **59**, 10493 (1999).
35. J. Maldacena, "*The Large N Limit of Superconformal Field Theories and Supergravity*", Adv. Theor. Math. Phys. **2**, 231 (1998).
36. S. S. Gubser, I. R. Klebanov, and A. M. Polyakov, "*Gauge Theory Correlators from Non-Critical String Theory*", Phys. Lett. B **428**, 115 (1998).
37. Edward Witten, "*Anti De Sitter Space And Holography*", Adv. Theor. Math. Phys. **2**, 253 (1998).
38. O. Aharony, S. S. Gubser, J. M. Maldacena, H. Ooguri, and Y. Oz, "*Large N Field Theories, String Theory and Gravity*", Phys. Rept. **323**,183 (2000).
39. Shinsei Ryu and Tadashi Takayanagi, "*Holographic Derivation of Entanglement Entropy from the anti-de Sitter Space/Conformal Field Theory Correspondence*", Phys. Rev. Lett. **96**, 181602 (2006)
40. Shinsei Ryu and Tadashi Takayanagi, "*Aspects of Holographic Entanglement Entropy*", JHEP 0608:45,2006.
41. Tatsuma Nishioka, Shinsei Ryu, and Tadashi Takayanagi, "*Holographic Entanglement Entropy: An Overview*", J. Phys. A **42**, 504008 (2009).
42. Mans Henningson and Kostas Skenderis, "*The Holographic Weyl Anomaly*", JHEP **9807**, 023 (1998).
43. Hiroshi Ueda and Tomotoshi Nishino, "*Hyperbolic Deformation on Quantum Lattice Hamiltonians*", J. Phys. Soc. Jpn. **78**, 014001 (2009).
44. Toshiya Hikihara and Tomotoshi Nishino, "*Connecting distant ends of one-dimensional critical systems by a sine-square deformation*", Phys. Rev. B **83**, 060414 (2011).
45. Isao Maruyama, Hosho Katsura, and Toshiya Hikihara, "*Sine-square deformation of free fermion systems in one and higher dimensions*", Phys. Rev. B **84**, 165132 (2011).
46. Hosho Katsura, "*Sine-square deformation of solvable spin chains and conformal field theories*", arXiv:1110.2459.
47. F. Verstraete, D. Porras, and J. I. Cirac, "*Density matrix Renormalization Group and Periodic Boundary Conditions: A Quantum Information Perspective*", Phys. Rev. Lett. **93**, 227205 (2004).
48. F. Verstraete and J. I. Cirac, "*Renormalization algorithms for Quantum-Many Body Systems in two and higher dimensions*", arXiv:cond-mat/0407066.
49. F. Verstraete, J. I. Cirac, and V. Murg, "*Matrix Product States, Projected Entanglet Pair States, and variational renormalization group methods for quantum spin systems*", Adv. Phys. **57**, 143 (2008).
50. A. Isacsson and O. F. Syljuåsen, "*Variational treatment of the Shastry-Sutherland antiferromagnet using projected entangled pair states*", Phys. Rev. E **74**, 026701 (2006).

51. V. Murg, F. Verstraete, and J. I. Cirac, "*Variational study of hard-core bosons in a two-dimensional optical lattice using projected entangled pair states*", Phys. Rev. A **75**, 033605 (2007).
52. V. Murg, F. Verstraete, and J. I. Cirac, "*Exploring frustrated spin-systems using Projected Entangled Pair States (PEPS)*", Phys. Rev. B **79**, 195119 (2009).
53. G. Vidal, "*Entanglement Renormalization*", Phys. Rev. Lett. **99**, 220405 (2007).
54. G. Vidal, "*Class of Quantum Many-Body States That Can Be Efficiently Simulated*", Phys. Rev. Lett. **101**, 110501 (2008).
55. G. Evenbly and G. Vidal, "*Algorithms for entanglement renormalization*", Phys. Rev. **B79**, 144108 (2009).
56. G. Evenbly and G. Vidal, "*Entanglement Renormalization in Two Spatial Dimension*", Phys. Rev. Lett. **102**, 180406 (2009).
57. Robert N. C. Pfeifer, Glen Evenbly, and Guifré Vidal, "*Entanglement renormalization, scale invariance, and quantum criticality*", Phys. Rev. A **79**, 040301(R)(2009).
58. Philippe Corboz and Guifré Vidal, "*Fermionic multiscale entanglement renormalization*", Phys. Rev. B **80**, 165129 (2009).
59. G. Evenbly, R. N. C. Pfeifer, V. Picó, S. Iblisdir, l. Tagliacozzo, I. P. McCulloch, and G. Vidal, "*Boundary critical phenomena with entanglement renormalization*", Phys. Rev. B **82**, 161107 (2010).
60. P. Silvi, V. Giovannetti, P. Calabrese, G. E. Santoro, and R. Fazio, "*Entanglement renormalization and boundary critical phenomena*", J. Stat. Mech. (2010) L03001.
61. Brian Swingle, "*Entanglement Renormalization and Holography*", arXiv:0905.1317.
62. G. Evenbly and G. Vidal, "*Tensor network states and geometry*", arXiv:1106.1082.
63. J. Ignacio Cirac, Didier Poilblanc, Nobert Schuch, and Frank Verstraete, "*Entanglement spectrum and boundary theories with projected entangled-pair states*", Phys. Rev. **83**, 245134 (2011).
64. Jie Lou, Shu Tanaka, Hosho Katsura, and Naoki Kawashima, "*Entanglement Spectra of the 2D AKLT Model: VBS/CFT Correspondence*", arXiv:1107.3888.
65. Masuo Suzuki, "*Relationship between d-Dimensional Quantal Spin Systems and $(d+1)$-Dimensional Ising Systems*", Prog. Theor. Phys. **56**, 1454 (1976).
66. Hiroaki Matsueda, "*Entanglement Entropy and Entanglement Spectrum for Two-Dimensional Classical Spin Configuration*", published in Phys. Rev. E.
67. Hiroaki Matsueda, "*Renormalization Group and Curved Spacetime*", arXiv:1106.5624.

FROM CLASSICAL NEURAL NETWORKS TO QUANTUM NEURAL NETWORKS

B. TIROZZI

Department of Physics, Rome University "La Sapienza"

First I give a brief description of the classical Hopfield model introducing the fundamental concepts of patterns, retrieval, pattern recognition, neural dynamics, capacity and describe the fundamental results obtained in this field by Amit, Gutfreund and Sompolinsky,[1] using the non rigorous method of replica and the rigorous version given by Pastur, Shcherbina, Tirozzi[2] using the cavity method. Then I give a formulation of the theory of Quantum Neural Networks (QNN) in terms of the XY model with Hebbian interaction. The problem of retrieval and storage is discussed. The retrieval states are the states of the minimum energy. I apply the estimates found by Lieb[3] which give lower and upper bound of the free-energy and expectation of the observables of the quantum model. I discuss also some experiment and the search of ground state using Monte Carlo Dynamics applied to the equivalent classical two dimensional Ising model constructed by Suzuki et al.[6] At the end there is a list of open problems

Keywords: Retrieval; Critical capacity; Qubits.

1. The Hopfield model

The model which has clearer statements about memory, retrieval and storage of patterns is the Hopfield model. The Hopfield model has also the property of associative memory. For associative memory we mean the capacity that we have to recover an entire information starting from a partial knowledge of it. The Hopfield model also has a precise way to define the capacity of some working memory to store information and it is interesting because its evolution resembles the dynamic of real neurones. Let us consider a system of N formal neurons, by formal we mean that we represent neurones with a simple variable $\sigma \in (-1, 1)$. Any neuron i is associated with a variable of this type σ_i. The value $+1$ is assigned to the neuron i if the neuron is active and -1 if it is not active. The activity of a neuron is a complicate process in which are involved many structures, called channels, which stay on the surface membrane of the neuron. Various species of charged ions pass through these channels determining a change of the electric potential on the surface of neuronal cell. These changes are very rapid and in a time less than 1 ms the electric potential reaches a maximum value and returns to the rest value, an e.m. wave goes out from the cell and the potential of the connected cell changes. This is a spike and the membrane of other neurons also make spikes when their potential becomes larger than a certain threshold θ. The scheme is the following.

State variables:
$$\sigma_i = \pm 1$$
$$\underline{\sigma} \equiv (\sigma_1, .., \sigma_N)$$

Neural dynamics:
$$\underline{\sigma}(t) \Rightarrow \underline{\sigma}(t+1)$$
$$\sigma_i(t+1) = \text{sign}(\sum_{j \neq i} J_{ij}\sigma_j(t) - \theta)$$

J_{ij}: synaptic weights
Patterns:
$$\xi \equiv \xi_i^\mu$$
$$\mu = 1, .., P, \quad i = 1, ..N$$

ξ_i^μ i.i.d.r.v. with value ± 1, $\Pr(1/2, 1/2)$ pattern $\underline{\xi}^\mu \equiv (\xi_1^\mu, ..., \xi_N^\mu)$.
μ index of pattern
Retrieval, associative memory
$$\sigma_i(0) = \epsilon_i \xi_i^\mu$$

$\epsilon_i \in (-1, 1)$, $i = 1, ..., N$, i.i.d.r.v. $\Pr(\epsilon_i = 1) = 1 - q$, $\Pr(\epsilon_i = -1) = q$, $q \leq \frac{1}{2}$

$$\underline{\sigma}(0) \longrightarrow \underline{\sigma}(1).... \longrightarrow \underline{\xi}^\mu$$

Hebb's rule
$$J_{ij} = \frac{1}{N} \sum_{\mu=1}^{P} \xi_i^\mu \xi_j^\mu$$

Overlap parameter

$$m^\mu = \frac{1}{N} \sum_{i=1}^{N} \xi_i^\mu \sigma_i$$

$\mu = 1, ...P.$

$$m^\mu = 1$$

retrieval of the pattern μ,

$$m^\mu = O(\frac{1}{\sqrt{N}})$$

if $\underline{\sigma}$ independent from $\underline{\xi}$
Capacity

$$\alpha = \frac{P}{N}$$

Hamiltonian

$$H(\underline{\sigma}, \xi) = -\frac{1}{2} \sum_{i \neq j} \frac{1}{N} \sum_{\mu=1}^{P} \xi_i^\mu \xi_j^\mu \sigma_i \sigma_j$$

Retrieval states $\longrightarrow$ local minima of H.

$$H(\underline{\sigma}, \xi) = \frac{P}{2} - \frac{N}{2} \sum_\mu (m^\mu)^2$$

$$H(\underline{\xi}^1, \xi) = \frac{P}{2} - \frac{N}{2} - \frac{P}{2} = -\frac{N}{2}$$

$$H(\underline{\sigma}, \xi) = 0$$

if $\underline{\sigma}$ independent from $\underline{\xi}$
$\longrightarrow$ system with P minima degenerate (equal values).
Monte Carlo Dynamic

$$\underline{\sigma} \longrightarrow \underline{\sigma}'$$

if $\Delta H(\underline{\sigma}, \xi) = H(\underline{\sigma}', \xi) - H(\underline{\sigma}, \xi) \leq 0)$
if $\Delta H(\underline{\sigma}, \xi) \geq 0)$ the transition happens with probability

$$\Pr \sim e^{-\beta \Delta H(\underline{\sigma}, \xi)}$$

The change of configuration is usually done for one spin, so $\underline{\sigma}' = (-\sigma_1, ..., \sigma_N)$ and then σ_2 is changed and so on. The neural dynamic brings the state in a retrieval state also the M.C, because :

$$\lim_{\beta \to \infty} \underline{\sigma}^{\text{Monte Carlo}}(t) \to \underline{\sigma}^{\text{Neural Dynamics}}(t)$$

Gibbs measure:
$$p(\underline{\sigma},\xi) = \frac{e^{-\beta H(\underline{\sigma},\xi)}}{\sum_{\underline{\sigma}} e^{-\beta H(\underline{\sigma},\xi)}} \tag{1}$$

with

$$H(\underline{\sigma},\xi) = -\frac{1}{2}\sum_{i\neq j}\frac{1}{N}\sum_{\mu=1}^{P}\xi_i^\mu \xi_j^\mu \sigma_i \sigma_j - h\sum_i \xi_i^\mu \sigma_i \tag{2}$$

The Gibbs measure is invariant with respect to M.C., it satisfies a stronger conditions, the principle of detailed balance:

$$p(\underline{\sigma},\xi)\Pr(\underline{\sigma}\longrightarrow\underline{\sigma}') = p(\underline{\sigma}',\xi)\Pr(\underline{\sigma}'\longrightarrow\underline{\sigma})$$

Ergodicity of M.C. dynamic

$$\sigma_i(t) \longrightarrow <\sigma_i> \tag{3}$$

where $<f(\underline{\sigma})> = \sum_{\underline{\sigma}} p(\underline{\sigma},\xi)f(\underline{\sigma})$ is the average with respect to the Gibbs measure.

Other definitions of overlap parameters

$$m^\mu = \frac{1}{N}\sum_i \sigma_i(t)\xi_i^\mu \tag{4}$$

where $\sigma_i(t)$ is the trajectory generated by M.C. dynamic starting from $\sigma_i(0)$. Using the ergodicity of M.C.

$$m_N^\mu = \frac{1}{N}\sum_i \xi_i^\mu <\sigma_i> \tag{5}$$

Partition function:

$$Z_N(\beta,\xi) \equiv \sum_{\underline{\sigma}} e^{-\beta H(\underline{\sigma},\xi)} \tag{6}$$

Free-energy:

$$f_N(\beta,\xi) = -\frac{1}{\beta N}\log Z_N(\beta,\xi) \tag{7}$$

Self-averaging. f_N, Z_N, m_N^μ are all functions of the r.v. ξ. We want to have retrieval independent from the choice of the ξ, we introduce then the property of self-averaging, in the limit $N\to\infty$

$$<<(g_N(\xi)-<<g_N(\xi)>>)^2>>\to 0 \tag{8}$$

where $<<>>$ is the average with respect to ξ

$$<<f(\xi))>> = \frac{1}{2^{PN}}\prod_{i=1,..,N,\mu=1,..,P}\sum_{\xi_i^\mu=\pm 1} f(\xi)$$

Solution of the model. P, number of patterns, finite. Applying saddle-point technique and using a simple argument of self-averaging we get for the free-energy, in the $N \to \infty$ limit:

$$f(\beta, h) = \frac{1}{2} \sum_{\mu=1}^{P} (m^\mu)^2 - \frac{1}{\beta} << \log 2 \cosh \beta (\sum_{\mu=1}^{P} m^\mu \xi_1^\mu + h\xi_1^\mu) >> \qquad (9)$$

where m^μ satisfies the saddle-point equation

$$\frac{\partial f}{\partial m^\mu} = 0 \Longrightarrow m^\mu = << \xi_1^\mu \tanh \beta (\sum_{\mu=1}^{P} m^\mu \xi_1^\mu + h\xi_1^\mu) >> \qquad (10)$$

We will see that the ground state energy of this model is degenerate with P minima of ferromagnetic type, all corresponding to memory states.

$P \to \infty$ with $N \to \infty$, α fixed.

Hamiltonian, s number of condensing patterns:

$$H(\underline{\sigma}, \xi) = -\frac{1}{2} \sum_{i \neq j} \frac{1}{N} \sum_{\mu=1}^{P} \xi_i^\mu \xi_j^\mu \sigma_i \sigma_j - \sum_{\mu=1}^{s} h^\mu \sum_i \xi_i^\mu \sigma_i \qquad (11)$$

Other order parameters:

Edward-Anderson (spin-glass) order parameter q_N:

$$q_N = \frac{1}{N} \sum_i <\sigma_i>^2 \qquad (12)$$

Non-condensing patterns order parameter r_N

$$r_N = \frac{1}{\alpha} \sum_{\nu=s+1}^{P} (m^\nu)^2 \qquad (13)$$

$$f(\beta, q, r, m^\mu) = \lim_{N \to \infty, P \to \infty, \alpha = \frac{P}{N} \text{ fixed}} -\frac{1}{\beta N} << \log \sum_{\underline{\sigma}} e^{-\beta H(\underline{\sigma}, h, \xi)} >>$$
$$\qquad (14)$$

Replica method.[1]

$$<< \log \sum_{\underline{\sigma}} e^{-\beta H(\underline{\sigma}, h, \xi)} >> = << \lim_{x \to 0} \frac{(\sum_{\underline{\sigma}} e^{-\beta H(\underline{\sigma}, h, \xi)})^x - 1}{x} >>$$

$$= << \lim_{n \to 0} \frac{(\sum_{\underline{\sigma}} e^{-\beta H(\underline{\sigma}, h, \xi)})^n - 1}{n} >> \qquad (15)$$

n integer, n replicas, $\lim n \to 0$ not rigorous

$$(\sum_{\underline{\sigma}} e^{-\beta H(\underline{\sigma}, h, \xi)})^n = \sum_{\underline{\sigma}^1, \ldots, \underline{\sigma}^n} e^{-\beta \sum_{a=1}^{n} H(\underline{\sigma}^a, h, \xi)}$$

n order parameters, q_a, r_a, m_a^μ, $a = 1,..,n$, Replica symmetry $q_a = q$, $r_a = r$, $m_a^\mu = m^\mu$.

Asymptotic free-energy, s number of condensing patterns

$$f = \frac{\alpha}{2} + \frac{1}{2}\sum_{\nu=1}^{s}(m^\nu)^2 + \frac{\alpha\beta r(1-q)}{2} + \frac{\alpha}{2\beta}[\ln(1-\beta(1-q)) - \frac{\beta q}{1-\beta(1-q)}]$$

$$-\frac{1}{\beta}\int\frac{dz}{\sqrt{2\pi}}e^{-\frac{z^2}{2}} << \log[2\cosh\beta(\sqrt{\alpha r}z + \sum_{\nu=1}^{s}(m^\nu + h^\nu)\xi_1^\nu)] >>$$

Saddle-point equations

$$\frac{\partial f}{\partial m^\nu} = 0 \Longrightarrow m^\nu = \int\frac{dz}{\sqrt{2\pi}}e^{-\frac{z^2}{2}} << \xi^\nu \tanh\beta(\sqrt{\alpha r}z + \sum_{\nu=1}^{s}(m^\nu + h^\nu)\xi_1^\nu) >>$$
(16)

$$\frac{\partial f}{\partial r} = 0 \Longrightarrow q = \int\frac{dz}{\sqrt{2\pi}}e^{-\frac{z^2}{2}} << \tanh^2\beta(\sqrt{\alpha r}z + \sum_{\nu=1}^{s}(m^\nu + h^\nu)\xi_1^\nu) >>$$
(17)

$$\frac{\partial f}{\partial q} = 0 \Longrightarrow r = \frac{q}{(1-\beta(1-q))^2}$$
(18)

Limit $\beta \to \infty$, $\alpha_c = 0.138$. The discussion about ground state energy will be given after.

Rigorous cavity method[2]

Proof of self-averaging properties

Self-averaging of the free-energy:

$$<< (f_N(\xi,\beta) - << f_N(\xi,\beta) >>)^2 >> \to 0, N \to \infty$$
(19)

$$<< (m_N^\mu(\xi,\beta) - << m_N^\mu(\xi,\beta) >>)^2 >> \to 0, N \to \infty$$

if $<< (q_N(\xi,\beta) - << q_N(\xi,\beta) >>)^2 >> \to 0, N \to \infty \Longrightarrow$

$$\Longrightarrow << (r_N(\xi,\beta) - << r_N(\xi,\beta) >>)^2 >> \to 0, N \to \infty$$

Hamiltonian. For the cavity method is much more complicated.

$$H(\underline{\sigma},\xi,\underline{\gamma},\underline{h},\epsilon_1,\epsilon_2) = -\frac{1}{2}\sum_{i\neq j}\frac{1}{N}\sum_{\mu=1}^{P}\xi_i^\mu\xi_j^\mu\sigma_i\sigma_j$$

$$-\sum_{\mu=1}^{s}\epsilon^\mu\sum_i\xi_i^\mu\sigma_i - \epsilon_1\sum_i h_i\sigma_i - \epsilon_2\sum_{\mu=s+1}^{P}\gamma^\mu t^\mu$$

where

- $\underline{\gamma} = \gamma^{s+1},..\gamma^P$ is a vector of gaussian random variables with zero mean and average 1.
- $\underline{h} = h_1,..,h_N$ is a vector of gaussian random variables with zero mean and average 1, independent from γ^μ.
- $\epsilon_1, \epsilon_2, \epsilon^\mu$ are given constants which go to zero at the end.
- $t^\mu = \frac{1}{\sqrt{N}}\sum_i \xi_i^\mu \sigma_i$

Main Result. Applying the cavity method is possible to show that, if q is s.a. then the saddle-point equations hold.

$$m^\nu = \int \frac{dz}{\sqrt{2\pi}} e^{-\frac{z^2}{2}} << \xi^\nu \tanh \beta(\sqrt{\alpha r}z + \sum_{\nu=1}^{s}(m^\nu + h^\nu)\xi_1^\nu) \quad (20)$$

$$q = \int \frac{dz}{\sqrt{2\pi}} e^{-\frac{z^2}{2}} << \tanh^2 \beta(\sqrt{\alpha r}z + \sum_{\nu=1}^{s}(m^\nu + h^\nu)\xi_1^\nu) >> \quad (21)$$

$$r = \frac{q}{(1 - \beta(1-q))^2} \quad (22)$$

2. Estimates of the observables for quantum spin systems

In order to simulate the retrieval process for quantum neural networks we have to find the ground states and show that the minima are states coinciding with the patterns. In the classical neural networks this can be done by means of the Monte Carlo dynamic which, as it was shown before, brings the states of the network to the retrieval of the patterns. The analogous dynamic for quantum spin systems is the Quantum Monte Carlo. But it would be extremely useful to have and estimate from above and below of the ground state energy and the expectation of the overlap parameters. In this section I show the useful estimates made by Lieb[3] which express the quantum expectations in terms of expectations of special functionals of classical spins. This idea having been introduced in the papers.[4,5] This approach is similar to the construction of Suzuki[6] which shows the equivalence of the one dimensional generic XY model with a special type of two dimensional Ising system of classical spins. This simulation[8] is different from the Quantum Monte Carlo and has been intensively used for translationally invariant system[7] although it can be applied also in the case of random interaction.

Let Σ be the unit sphere in 3d dimensions

$$\Sigma = \{(x,y,z)|x^2+y^2+z^2=1\}$$

$L^2(\Sigma)$ the space of square integrable functions on Σ with the usual measure

(1) $\Omega = (\theta,\phi), 0 \leq \theta \leq \pi, 0 \leq \phi, \leq 2\pi$
(2) $d\Omega = \sin\theta d\theta d\phi$
(3) $x = \sin\theta\cos\phi, y = \sin\theta\sin\phi, z = \cos\theta$

I will make use of spins with $J = \frac{1}{2}, \frac{3}{2}$ for getting useful inequalities.
$\hat{\sigma}^x, \hat{\sigma}^y, \hat{\sigma}^z$ be the Pauli matrices
$\hat{\sigma}^{\pm} = \hat{\sigma}^x \pm i\hat{\sigma}^y$ be the creation and annichilation operators.
If one takes the "'spin up'" state $|J> \in C^{2J+1}$, $\hat{\sigma}^z|J> = J|J>$, then the generic Bloch state $|\Omega> \in C^{2J+1}$

$$|\Omega> = \sum_{M=-J}^{J} (\binom{2J}{M+J})^{\frac{1}{2}} (\cos\frac{\theta}{2})^{J+M} (\sin\frac{\theta}{2})^{[J-M]\phi}|M> \quad (23)$$

This state is a generalization of the usual spin 1/2 state

$$|\psi> = \cos\frac{\theta}{2}|0> + e^{i\phi}\sin\frac{\theta}{2}|1>$$

to the generic spin J.

The overlap among $|\Omega>$ states is given by

$$(\Omega',\Omega) \equiv <\Omega'|\Omega> = \quad (24)$$

$$[\cos\frac{\theta}{2}\cos\frac{\theta'}{2} + \exp i(\phi-\phi')\sin\frac{\theta}{2}\sin\frac{\theta'}{2}]^{2J} \quad (25)$$

For any operator $\hat{A}$ acting on the space of states C^{2J+1} it is possible to find two functions $G(\Omega)$ and $g(\Omega)$ such that

$$\hat{A} = \frac{2J+1}{4\pi} \int_\Sigma d\Omega G(\Omega)|\Omega><\Omega| \quad (26)$$

$$g(\Omega) = <\Omega|\hat{A}|\Omega> \quad (27)$$

The functions $g(\Omega)$ and $G(\omega)$ are given in[3] for elementary spin operators

$$\hat{\sigma}_{x,y,z} = \frac{2J+1}{4\pi} \int_\Sigma d\Omega G_{x,y,z}(\Omega)|\Omega><\Omega| \qquad (28)$$

$$g(\Omega)_{x,y,z} = <\Omega|\hat{\sigma}_{x,y,z}|\Omega> \qquad (29)$$

Operator	$g(\Omega)$	$G(\Omega)$
$\hat{\sigma}_z$	$J\cos\theta$	$(J+1)\cos\theta$
$\hat{\sigma}_x$	$J\sin\theta\cos\phi$	$(J+1)\sin\theta\cos\phi$
$\hat{\sigma}_y$	$J\sin\theta\sin\phi$	$(J+1)\sin\theta\sin\phi$

In order to compute the average values of the quantum observables it is necessary to evaluate the quantum partition function

$$Z^Q = \alpha_N \text{Tr} e^{-\beta \hat{H}} \qquad (30)$$

where $\alpha_N = \prod_{i=1}^N (2J+1)^{-N}$ is a normalization factor, the Hilbert space is

$$\mathcal{H}_N = \otimes_{i=1}^N \mathcal{H}_i = \otimes_{i=1}^N C^{2J+1}$$

The complete normalized set of states $|\Omega_N>$ is

$$|\Omega_N> = \otimes_{i=1}^N |\Omega^i>$$

In this notation we can write

$$Z^Q = (4\pi)^{-N} \int d\Omega_N <\Omega_N|e^{-\beta\hat{H}}|\Omega_N>$$

Applying the Peierls-Bogoliubov's inequality $<\psi|e^X|\psi> \geq \exp<\psi|X|\psi>$ for any self-adjoint operator and any normalized $\psi \in \mathcal{H}_N$, we get

$$Z^Q \geq (4\pi)^{-N} \int d\Omega_N e^{-\beta<\Omega_N|\hat{H}|\Omega_N>} \equiv Z^C(J) \qquad (31)$$

Practically this formula gives a lower bound for Z^Q in terms of a classical partition function obtained by substituting in the Hamiltonian the operators $\hat{\sigma}^i_{x,y,z}$ and their products with the classical vector

$$J(\sin\theta^i \cos\phi^i, \sin\theta^i \sin\phi^i, \cos\theta^i)$$

and integrating over the angles. With some other inequalities on the trace of operators it is possible to derive the other useful inequality

$$Z^Q \leq (4\pi)^{-N} \int d\Omega_N e^{-\beta G(\Omega_N)} \equiv Z^C(J+1) \tag{32}$$

so we conclude with the useful bounds

$$Z^C(J) \leq Z^Q \leq Z^C(J+1) \tag{33}$$

with $J = 1/2$ and all the J in the two Hamiltonians being equal. I am going to apply these inequalities in the next section to the QNN Hamiltonian.

3. Multi-qubit systems and quantum neural networks

The general form of a N-qubit Hamiltonian which is useful for QNN can be written in the form:

$$\hat{H} = -\sum_{i=1}^{N}(\frac{\Delta_i}{2}\hat{\sigma}_i^x + \frac{\epsilon_i}{2}\hat{\sigma}_i^z) - \sum_{ij} J_{ij}\hat{\sigma}_i^z\hat{\sigma}_j^z \tag{34}$$

this is a Hamiltonian of the type of the XY model.

In the paper for quantum pattern recognition[9] the coefficients J_{ij} are chosen in analogy with the Hopfield model.

(1) J_{ij} is the Hebb interaction defined by means of the P patterns $\xi_i^\mu = \pm 1$, $\mu = 1, \ldots, P, i = 1, \ldots, N$, where the ξ_i^μ are the usual i.i.d r.v.

$$J_{ij} = \frac{1}{N}\sum_{\mu=1}^{P}\xi_i^\mu \xi_j^\mu \tag{35}$$

(2) The retrieval of the generic pattern in the classical Hopfield model can be obtained by adding a symmetry breaking term to the Hamiltonian

$$\sum_i \xi_i^\mu \hat{\sigma}_j^z \tag{36}$$

and so $\frac{\epsilon_i}{2} = \xi_i^\mu$ for the retrieval of the μ-th pattern.

The quantity to compute are the quantum partition function

$$Z^Q = \alpha_N \text{Tr} e^{-\beta \hat{H}} \tag{37}$$

and the quantum expectation of the overlap parameter

$$m^\mu = \frac{1}{N}\sum_i \xi_i^\mu \hat{\sigma}_i^z \tag{38}$$

$$<m^\mu>^Q = \frac{\text{Tr}\, m^\mu e^{-\beta \hat{H}}}{\text{Tr}\, e^{-\beta \hat{H}}}. \tag{39}$$

In order to get an estimate from above and below of Z^Q and $<m^\mu>^Q$ we apply the Peierls-Bogoliubov inequality ([3])

$$\lambda <A>^Q \geq f(\lambda) - f(0) \tag{40}$$

where

$$f(\lambda) = -\frac{1}{\beta} \ln \text{Tr}\, e^{-\beta(\hat{H}+\lambda A)} \tag{41}$$

with $\lambda > 0$ and A being a self-adjoint operator, which in our case is the overlap parameter m^μ. So we get

$$[f(0) - f(-\lambda)]/\lambda \geq <A>^Q \geq [f(\lambda) - f(0)]/\lambda \tag{42}$$

taking the limit $\beta \to \infty$ we obtain bounds on the quantum ground state energy

$$E_-^C \leq E^Q \leq E_+^C \tag{43}$$

where E^C is the classical ground state energy and the $-$ and the $+$ refers to the bound 32 and 31 respectively, we thus obtain the inequality

$$E^C(J) \geq E^Q \geq E^C(J+1) \tag{44}$$

I am going to use inequalities for the free-energy of the quantum system

$$f_N^Q(\Gamma, J) = -\frac{1}{\beta} \ln \text{Tr}\, e^{-\beta(\hat{H}+\lambda A)} \tag{45}$$

in order to estimate the ground state energy for $\beta \to \infty$. From the above arguments we have the inequalities

$$Z^C(\Gamma, J) \leq Z^Q(\Gamma, J) \leq Z^C(J+1)$$

$$f_N^C(\Gamma, J+1) \leq f_N^Q(\Gamma, J) \leq f_N^C(\Gamma, J)$$

where $Z^C(\Gamma, J)$ and $f_N^C(\Gamma, J)$ are the classical partition function and free-energy respectively.

$$Z_N^C(\Gamma, J) = \int \prod_{i=1}^{N} \sin\theta_i d\theta_i d\phi_i$$

$$e^{\frac{\beta\Lambda(s)J}{2} \sum_i \sin\theta_i \cos\phi_i + \frac{\beta}{2} \sum_{i\neq j} J_{ij} J^2 \cos\theta_i \cos\theta_j - \beta\Gamma \sum_i \xi_i^1 \cos\theta_i} \tag{46}$$

I will consider only the case P finite and $N \to \infty$, after some simple calculations we get for the classical free-energy

$$f_N^C = -\frac{1}{\beta N} \log Z^C(\Gamma, J) \tag{47}$$

$$= \sum_\mu \frac{1}{2}(m^\mu)^2 -$$

$$\frac{1}{\beta} <<\log \int du\, e^{-\frac{\beta}{2}J^2\alpha u^2 + \beta J^2 u \sum_\mu m^\mu \xi^\mu - \beta \Gamma u \xi^1 - \log(2\pi I_0(\frac{\beta\Lambda(s)}{2}\sqrt{1-u^2}))}>> \tag{48}$$

where the $<<.>>$ is the expectation with respect the ξ variables appearing in the formula, I_0 is the modified Bessel function of zero order and m^μ satisfies the saddle-point equation, $\frac{\partial f_N^C}{\partial m^\mu} = 0$

$$m^\nu = <<\frac{\int du\, e^{-\frac{\beta}{2}J^2\alpha u^2 + \beta J^2 u \sum_\mu m^\mu \xi^\mu - \beta \Gamma u \xi^1 - \log(2\pi I_0(\frac{\beta\Lambda(s)}{2}\sqrt{1-u^2}))} u J^2 \xi^\nu}{\int du\, e^{-\frac{\beta}{2}J^2\alpha u^2 + \beta J^2 u \sum_\mu m^\mu \xi^\mu - \beta \Gamma u \xi^1 - \log(2\pi I_0(\frac{\beta\Lambda(s)}{2}\sqrt{1-u^2}))}}>> \tag{49}$$

4. Estimates of the ground state energy of classical neural networks

In the introduction to classical neural networks we have seen that the retrieval states are the ones which minimize the hamiltonian of the system. If the $\lim \beta \to \infty$ is done in the formula for the free-energy then the ground state energy for the classical neural network is obtained. We give here a review of the estimates of the ground energy for the models which are important for our considerations.

4.1. Ising ferromagnetic model

The simplest system is the Ising ferromagnetic ($J > 0$)

$$H_N = -\frac{J}{2N} \sum_{i \neq j} \sigma_i \sigma_j - h \sum_i \sigma_i$$

The "overlap parameter" in this case is simply the magnetization

$$m = \frac{1}{N} \sum_i \sigma_i$$

The free-energy is

$$f(m) = \frac{1}{2}Jm^2 - \frac{1}{\beta}\log(2\cosh\beta(Jm - h))$$

In order to study phase transitions we have to send $h \to 0$ then m satisfies the saddle-point equation

$$m = \tanh J\beta m$$

for $\beta \to \infty$ $m \to \pm 1$ and this gives the value of the ground state energy is

$$E = (\frac{J}{2}m^2 - J|m|)N = -\frac{J}{2}N \qquad (50)$$

This simple calculation is instructive because it gives the idea of the structure of the ground states for the more complicate systems we are dealing with.

4.2. *Hopfield model with a finite number of patterns*

We have to study the limit $\beta \to \infty$ of the free-energy

$$f(\beta, h) = \frac{1}{2}\sum_\mu (m^\mu)^2 - \frac{1}{\beta} << \log 2(\cosh \beta(\sum_\mu m^\mu \xi_1^\mu + h\xi_1^1)) >>$$

with m^μ satisfying the saddle point-equation

$$m^\mu = << \xi_1^\mu \tanh \beta(\sum_\mu m^\mu \xi_1^\mu + h\xi_1^1)) >>$$

If we study the case of retrieval of only one pattern $m^\mu = m\delta_{\mu,1}$ and we send, as before $h \to 0$ we get:

$$f(\beta, 0) = \frac{1}{2}m^2 - \log 2 \cosh \beta m \qquad (51)$$

with m satisfying the saddle point equation

$$m = \tanh \beta m$$

i.e. we obtain the same equations of the ferromagnetic case and, in the limit $\beta \to \infty$ $m \to \pm 1$ and the asymptotic energy is $E = -\frac{1}{2}N$. But the difference with the ferromagnetic case is that the ground state is degenerate because there are P minima with the same value. The same degeneracy holds for $P \to \infty$, $(N \to \infty)$ and this makes the system not easy to work. Also for $P \to \infty$ the self-averaging will be more difficult to prove than this case.

4.3. *Hopfield model with finite capacity α*

Retrieval states

$$E = (-\frac{m^2}{2} + \frac{\alpha}{2}(1-r))N$$

$m \sim 1$, $r \sim 0$ (pure retrieval), real retrieval $m = 1 - \epsilon$, $\epsilon \sim 0.1 - 0.2$, $q \sim 0$.
for $\alpha = \alpha_c = 0.138$

$$E = -0.5014N$$

Spin glass state

$$r = (1 + \sqrt{\frac{2}{\pi\alpha}})^2$$

$$E_{SG} = (-\frac{1}{\pi} - \sqrt{\frac{2\alpha}{\pi}})N$$

for $\alpha = \alpha_c = 0.138$

$$E_{SG} = -0.615N$$

4.4. Quantum neural networks

We use the same argument for estimating and finding the bounds and asymptotic value of the free-energy and the ground state energy of the QNN. We apply the saddle-point to the integral appearing in the formula for the free-energy

$$\log \int du e^{-\frac{\beta}{2}J^2\alpha u^2 + \beta J^2 u \sum_\mu m^\mu \xi^\mu - \beta \Gamma u \xi^1}$$

$$\sim -\beta \min_u [\frac{1}{2}J^2\alpha u^2 - u(J^2 \sum_\mu m^\mu \xi^\mu - \beta \Gamma \xi^1)]$$

the minimum is obtained for

$$u_0 = \frac{1}{\alpha J^2}(J^2 \sum_\mu m^\mu \xi^\mu - \beta \Gamma \xi^1)$$

Substituting in the expression for the free-energy:

$$E^C(\Gamma, J) = \frac{1}{2}\sum_\mu (m^\mu)^2 + \frac{1}{2\alpha J^2} << (J^2 \sum_\mu m^\mu \xi^\mu - \Gamma \xi^1)^2 >> \quad (52)$$

$$= \frac{1}{2}\sum_\mu (m^\mu)^2 + \frac{1}{2\alpha J^2}(J^4 \sum_\mu (m^\mu)^2 + \Gamma^2 - 2\Gamma J^2 m^1) \quad (53)$$

Choosing a state of pure retrieval $m^\mu = m\delta_{\mu,1}$:

$$E^C(\Gamma, J) = \frac{1}{2}(1 + \frac{J^2}{\alpha})m^2 + \frac{\Gamma^2}{\alpha J^2} - 2\frac{\Gamma}{\alpha}m. \quad (54)$$

where m satisfies the saddle-point equation for the retrieval of a single pattern:

$$m = \left\langle\!\!\left\langle \frac{\int du \, e^{-\frac{\beta}{2}J^2\alpha u^2 + \beta\xi^1 u(J^2m-\Gamma)} u J^2 \xi^1}{\int du \, e^{-\frac{\beta}{2}J^2\alpha u^2 + \beta\xi^1 u(J^2m-\Gamma)}} \right\rangle\!\!\right\rangle \quad (55)$$

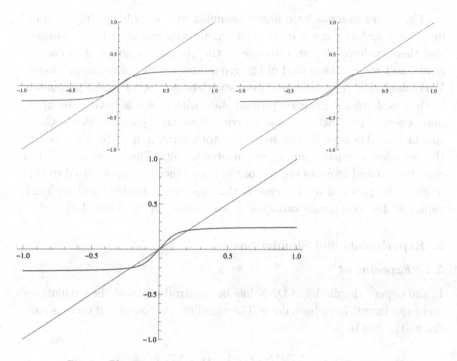

Fig. 1. Phase transition in the classical model, $\alpha = 0.1$, $\beta \sim 200$.

The numerical solution of this equation shows a first-order phase transition, in the figure 1, it is shown that the overlap parameter takes values different from zero at $\beta \sim 200$ for $\alpha \sim 0.1$ and it is zero for $\beta < 200$. It is possible also to find the value of α_c at zero temperature making the limit $\beta \to \infty$ in the equation 55. The point of minumum is

$$u_0 = \frac{(J^2m - \Gamma)\xi^1}{\alpha J^2}$$

Inserting this value in the asymptotic expansion for $\beta \to \infty$ in the equation 55

$$m = <<u_0 J^2 \xi^1>> \tag{56}$$

$$= \frac{(J^2 m - \Gamma)}{\alpha} + << \frac{\int_0^1 du\, e^{-\frac{\beta}{2}(u-u_0)^2 \alpha J^2}(u-u_0)}{\int_0^1 du\, e^{-\frac{\beta}{2}(u-u_0)^2 \alpha J^2}} >> \tag{57}$$

$$= \lim_{\Gamma \to 0} \to \frac{J^2 m}{\alpha} + \frac{1}{\sqrt{\beta \alpha}} [e^{-\frac{\beta}{2\alpha}(Jm)^2} - e^{-\frac{\beta}{2\alpha}(J)^2(1-\frac{m}{\alpha})^2}] \tag{58}$$

This expression has been found assuming $m < \alpha$. Solving this equation for β large and $m < \alpha$ the overlap parameter as a function of α is obtained and then we have also an estimate of the ground energy of the classical model and we get the bound of the ground energy of the quantum model. We remark that the saddle-point equation are symmetric in the parameter μ, thus we have a degenerate ground state with P states having the minimum energy. This fact makes more complicate the Quantum Monte Carlo simulations. These estimates and asymptotic formula have been derived in the hypothesis that the number of patterns is finite. The case $\alpha \neq 0$ is more complicated and requires the use of the cavity method[2] generalized to this model. The problem in this case is that instead of having spins with ± 1 values we have continuos variables u_i with values in the interval $(0,1)$.

5. Experiments and Simulations

5.1. *Experiment*

In the paper[9] the model of QNN has been introduced and the simulations and experiments have been done. The Hamiltonian considered there is similar to the one in (60)

$$\hat{H} = -\sum_{i=1}^{N} \left(\frac{\Delta_i}{2} \hat{\sigma}_i^x + \frac{\epsilon_i}{2} \hat{\sigma}_i^z \right) - \sum_{ij} J_{ij} \hat{\sigma}_i^z \hat{\sigma}_j^z \tag{59}$$

with the assumption that $N = 2$ for comparing with the experiments done on NMR systems. Thus the Hamiltonian is

$$\hat{H} = -\sum_{i=1}^{2} \Lambda(s) \frac{1}{2}(1 - \hat{\sigma}_i^x) - J_{12} \hat{\sigma}_1^z \hat{\sigma}_2^z - \Gamma \sum_{i=1}^{2} \xi_i^1 \hat{\sigma}_i^z \tag{60}$$

where the term multiplying Γ is the usual perturbation inserted in the Hamiltonian of the Hopfield model for retrieving the fist pattern ξ^1, thus Γ goes to zero for having the retrieval property. $\Lambda(s)$ is time-dependent decreasing function for the quantum adiabatic computation. In the experiment it is reduced linearly from $\Lambda(0) = A_{\max}$ to 0 in 50 ms a time shorter than the relaxation times T_1 and T_2 of the two $\frac{1}{2}$ nuclear spins of ^1H and ^{13}C. $T_1(^1H) = 1.6$ seconds, $T_1(^{13}C) = 2.7$ seconds, $T_2(^1H) = 130$ milliseconds, $T_2(^{13}C) = 60$ milliseconds. The experiments were performed at a temperature of $27°C$ using a Bruker AC 200 spectrometer operating at the Larmor frequency of 200 MHz for ^1H and 50 MHz for ^{13}C The term $\sum_{i=1}^{2} \Lambda(s)(\frac{1}{2}(1-\hat{\sigma}_i^x)$ is also introduced for generating the $\pi/2$ pulses used for measuring the states of the two qubits. The initial state is $|\psi> = \frac{1}{2}(|00> + |01> + |10> + |11>)$. J_{12} is set equal to $\frac{1}{2}(\xi_1^1\xi_2^1 + \xi_1^2\xi_2^2) \equiv w$ in agreement with the Hebb rule

$$w = \frac{1}{2}\sum_{\mu=1}^{2}\xi_1^\mu\xi_2^\mu = \frac{1}{2}(\xi_1^1\xi_2^1 + \xi_1^2\xi_2^2)$$

Thus $w = -1$ for the patterns $\xi^1 = (-1,1)$ and $\xi^2 = (1,-1)$ and $w = 1$ for the patterns $\xi^1 = (-1,-1)$ and $\xi^2 = (1,1)$. The perturbing Hamiltonian used in these experiments is not coinciding exactly with the usual Hopfield Hamiltonian because the perturbing term is slightly different, in fact it is given by

$$\Gamma(\overline{\xi_1}\hat{\sigma}_1^z + \overline{\xi_2}\hat{\sigma}_2^z)$$

where the pattern $\overline{\xi}$ have values $-1, 1$ and also 0. In the case of 0 it means absence of a bit and the 0 is considered as an incomplete information or wrong information. Thus the hamiltonian without the QAC term is

$$H_p = \hat{H} = -w\hat{\sigma}_1^z\hat{\sigma}_2^z + \Gamma(\overline{\xi_1}\hat{\sigma}_1^z + \overline{\xi_2}\hat{\sigma}_2^z)$$

The Hamiltonian of the NMR free evolution is

$$H_p' = 2\pi J\hat{\sigma}_H^z\hat{\sigma}_C^z + 2\pi\nu_H\hat{\sigma}_H^z + 2\pi\nu_C\hat{\sigma}_C^z$$

The offset frequencies are chosen according to the choice of the patterns $\nu_H = -w\Gamma J\overline{\xi_1}$ and $\nu_C = -w\Gamma J\overline{\xi_2}$, $J = 195$ Hz. Thus the choice $w = 1$ corresponds to the ground state while $w = -1$ is the highest energy state. The ν_H and ν_C where set equal to ± 100 Hz if $\overline{\xi_1} = \pm 1$ and 0 if $\overline{\xi_1} = 0$ respectively. The results of retrieval are found applying $\pi/2$ pulses and detecting the ^1H signal or ^{13}C signal. The result of the experiment depend on the choice of the perturbing patterns $\overline{\xi_1}, \overline{\xi_2}$ the component of this pattern can be ± 1 or 0. The output is summarized in the following table[9]

We see interesting results. If $w = -1$ the state of minimum energy is the one with opposite qubits and input pattern to the network is just the spin of ^1H and then the output state is the $+1$ spin of ^{13}C with the same qubit of ^1H. In other terms we give a pattern which is half of the whole

ξ_1	$\bar{\xi}_1$	$|\psi^{out}>$ for $w = -1$	$|\psi^{out}>$ for $w = +1$
-1	0	$|-1,1>$	$|-1,-1>$
1	0	$|1,-1>$	$|1,1>$
0	-1	$|1,-1>$	$|-1,-1>$
0	1	$|-1,1>$	$|1,1>$
0	0	$\frac{1}{\sqrt{2}}(|-1,1>+|1,-1>)$	$\frac{1}{\sqrt{2}}(|-1,-1>+|1,1>)$

pattern and we get the whole pattern. This retrieval is much stronger than the one in the classical Hopfield model where an input with 50% of bits wrong or absent would never give the whole pattern, this nice property is due to quantum effects and the fact that the retrieval state is the ground state of the Hamiltonian. Another important issue is that if the input is the null pattern then the network gives a combination of the two states of minimum energy for the coupling chosen. This effect is new altough in classical neural networks something similar could happen since the network con be in a spurious state which is made of states of spins belonging to different patterns. But in the classical case this combination is casual while in the quantum case is just the state of minima energy. Thus there are new effects but it is necessary to check if this phenomena are not limited to the case $N = 2$, in effect we have seen that N must go ∞ in order to have the retrieval for almost any set of patterns, i.e. for having the self-averaging property typical of the neural networks. Thus further theoretical and experimental work is needed in order to explore the effectiveness of the QNN.

5.2. *Simulations*

One possibility is the Quantum Monte Carlo, another is to use the equivalence established by Suzuki[6] and applied in many simulations.[7] The result is that the partition function of the quantum system:

$$\hat{H} = -\Lambda(s) \sum_{i=1}^{N} \hat{\sigma}_i^x - \Gamma \sum_i \xi_i^1 \hat{\sigma}_i^z - \sum_{ij} J_{ij} \hat{\sigma}_i^z \hat{\sigma}_j^z \quad (61)$$

is given by the two dimensional partition function of a classical Ising system of special form

$$Z^Q = \lim_{n \to \infty} \prod_{i=1,..,N, k=1,..,n}$$
$$\sum_{\sigma_{i,k}} e^{\sum_{i,j=1}^{N} \sum_{k=1}^{n} \frac{1}{n} J_{ij} \sigma_{i,k} \sigma_{j,k} + \frac{1}{2} \log \frac{n}{\beta \Lambda(s)} \sum_{i=1}^{N} \sum_{k=1}^{n} \sigma_{i,k} \sigma_{i,k+1} + \frac{\beta \Gamma}{n} \sum_{i=1}^{N} \xi_i^1 \sum_{k=1}^{n} \sigma_{ik}}$$

6. Open Problems

- Complete the estimates from above and below of the ground energy of the quantum model for P finite and $N \to \infty$.
- Compute exactly the free-energy in the above case using Bogoliubov's approach similar to the Curie-Weiss model.
- Use replica and/or cavity model for the estimates in the case α finite.
- Make the experiments with more qubits with Liquid State NMR or other apparatus.
- Simulation of the system with the two dimensional equivalent Ising model or with Quantum Monte Carlo.

References

1. D. Amit, H. Gutfreund, H. Sompolinsky, Ann. of Phys., 173, 30-67, 1987.
2. L. Pastur, M. Shcherbina, B. Tirozzi, J. Stat. Phys. 74 5/6, 1167-1183, 1994.
3. E. Lieb, The Classical Limit of Quantum Spin Systems, Communication Math. Phys., 31, 327-340, 1973.
4. F. T. Arecchi, E. Courtens, R. Gilmore, H. Thomas, Phys. Rev. A6, 2211-2237, 1972.
5. J. M. Radcliffe, J. Phys. A, 4, 313-323, 1971.
6. M. Suzuki, Relationship between d-dimensional quantal spin systems and (d+1)-dimensional Ising systems, Progress of Theoretical Physics, V. 56, 5, 1976.
7. Quantum Monte Carlo Methods in equilibrium and nonequlibrium systems, Editor: M. Suzuki, Springer Series in Solid-State Sciences 74, Springer-Verlag 1987.
8. M. Suzuki, S. Miyashita and A. Kuroda, Monte Carlo Simulation of Quantum Spin Systems I., Progress of Theoretical Physics, Vol. 58, N 5, p. 1377-1387, 1977.
9. R. Neigovzen, J. Neves, R. Sollacher, S. Glaser, Quantum pattern recognition with liquid state NMR, Phys. Rev. A 79, 042321, 2009.

ANALYSIS OF QUANTUM MONTE CARLO DYNAMICS IN INFINITE-RANGE ISING SPIN SYSTEMS: THEORY AND ITS POSSIBLE APPLICATIONS

JUN-ICHI INOUE

Graduate School of Information Science and Technology, Hokkaido University
N14-W9, Kita-Ku, Sapporo 060-0809, Japan
*E-mail: j_inoue@complex.ist.hokudai.ac.jp, jinoue@cb4.so-net.ne.jp
http://chaosweb.complex.eng.hokudai.ac.jp/~j_inoue/

In terms of the stochastic process of a quantum-mechanical variant of Markov chain Monte Carlo method based on the Suzuki-Trotter decomposition, we analytically derive deterministic flows of order parameters such as magnetization in infinite-range (a mean-field like) quantum spin systems. Under the static approximation, differential equations with respect to order parameters are explicitly obtained from the Master equation that describes the microscopic-law in the corresponding classical system. We discuss several possible applications of our approach to several research topics, say, image processing and neural networks. This paper is written as a self-review of two papers[1,2] for *Symposium on Interface between Quantum Information and Statistical Physics* at Kinki University in Osaka, Japan.

Keywords: Quantum dynamics, Quantum Monte Carlo method, Infinite-range model, Probabilistic information processing, Image restoration, Neural networks, Associative memories.

1. Introduction

In various research fields including information science or economics, the Markov chain Monte Carlo method (the MCMC) has been widely used to calculate various physical quantities (expectations) or to construct marginal distributions by sampling the important states that contribute effectively to the quantities.[3] Generally speaking, the MCMC needs a long time to wait until the Markovian stochastic process starts to generate the microscopic states from a well-approximated distribution. Especially, for some classes of probabilistic models which are categorized in the so-called random spin systems including *spin glasses*,[4–6] the time consuming is sometimes very serious problem to make attempt to proceed the desired information processing (*e.g.* finding the ground state approximately) within a realistic computational time. However, even for such cases, various effective improvements based on several important concepts have been proposed and succeeded in carrying out the numerical calculations.[7,8] From the view point of statistical physics, transitions between microscopic states are controlled by a specific parameter, namely, 'temperature' of the system and by cooling the temperature slowly enough during the Markovian process, one can get the lowest energy states efficiently. This type of optimization tool based on 'thermal fluctuation' is referred to as *simulated annealing*.[9,10]

Recently, the simulated annealing has been extended to the quantum-mechanical version. This 'new type of simulated annealing' called as *quantum annealing*[11–15] is based on the adiabatic theorem of the quantum system that evolves according to Schrödinger equations. To use the quantum annealing, or more generally, to utilize the quantum fluctuation for combinatorial optimization problems or massive information processing (*e.g.* image restoration or error-correcting codes), the approach by solving the Schrödinger equations is apparently limited (almost impossible) and we should look for another 'shortcut' to simulate the quantum systems.

As the most effective and efficient way, the quantum Monte Carlo method[16] was established. The method is based on the following Suzuki-Trotter decomposition for non-commutative two operators $\mathcal{A}$ and $\mathcal{B}$:

$$\mathrm{tr}\exp\left(\mathcal{A}+\mathcal{B}\right) = \lim_{M \to \infty} \mathrm{tr}\left(\exp\left(\frac{\mathcal{A}}{M}\right)\exp\left(\frac{\mathcal{B}}{M}\right)\right)^M \quad (1)$$

Namely, the d-dimensional quantum system is mapped to the corresponding $(d+1)$-dimensional classical spin systems. This approach is very powerful and a lot of researches have succeeded in exploring the quantum phases in strongly correlated quantum systems. However, when we simulate the quantum system at zero temperature in which quantum effect is essential, we encounter some technical difficulties although several sophisticated algorithms were proposed.[17] From the view point of information science, it is very informative for us to evaluate the process of information processing at zero temperature, and if one seeks to utilize the quantum fluctuation to solve the problems, we should use the 'zero-temperature dynamics'. For this purpose, it seems that we need some tractable 'bench mark tests' to investigate the 'dynamical process' of information processing at zero temperature.

In classical system, Coolen and Ruijgrok[18] proposed a way to derive the differential equations with respect to order-parameters of the system from the microscopic master equations. They dealt with the so-called *Hopfield model*[19,20] as an associative memory in which a finite number of patterns are embedded. The procedure was extended by Coolen and Sherrington,[21] Coolen, Laughton and Sherrington[22] to more complicated spin systems categorized in the infinite-range (mean-field like) models including the Sherrington-Kirkpatrick spin glasses.[23] The so-called dynamical replica theory (DRT) was now well-established as a strong approach to investigate the dynamics in the classical disordered spin systems. On the considering the matter, it seems to be important for us to extend their approach to quantum systems evolving stochastically according to the quantum Monte Carlo dynamics.

In the reference,[1] the present author proposed a formulation of deriving differential equations of macroscopic order parameters from the microscopic master equations in terms of the Suzuk-Trotter decomposition. The formulation was applied to the ferromagnetic infinite-range Ising model and image restoration problem as a sort of random field Ising models. The method was examined for the problem of pattern recalling process in the so-called Hopfield model.[2] In this self-review paper, we explain the details of these references[1,2] as follows.

In terms of the stochastic process of quantum-mechanical version of Markov chain Monte Carlo method, we analytically derive macroscopically deterministic flow equations of order parameters such as spontaneous magnetization in infinite-range ($d(=\infty)$-dimensional) quantum spin systems. By means of the Trotter decomposition, we consider the transition probability of Glauber-type dynamics of microscopic states for the corresponding $(d+1)$-dimensional classical system. Under the static approximation, differential equations with respect to macroscopic order parameters are explicitly obtained from the master equation that describes the microscopic-law. In the steady state, we show that the equations are identical to the saddle point equations (equations of states) for the equilibrium state of the same system. We easily find that the equation for the dynamical Ising model is recovered in the classical limit. We also discuss several possible applications of our approach to several research areas, namely, information processing and neural networks.

2. The model system and formulation

In this section, we derive the differential equations with respect to several order parameters for a simplest quantum spin system, namely, a class of the infinite range transverse Ising model[30,31] described by the following

Hamiltonian:

$$H = -\frac{1}{N}\sum_{i,j=1}^{N} J_{ij}\sigma_i^z\sigma_j^z - h\sum_{i=1}^{N}\tau_i\sigma_i^z - \Gamma\sum_{i=1}^{N}\sigma_i^x \quad (2)$$

where σ_i^z and σ_i^x denote the Pauli matrices given by

$$\sigma_i^z = \begin{pmatrix} 1 & 0 \\ 0 & -1 \end{pmatrix}, \quad \sigma_i^x = \begin{pmatrix} 0 & 1 \\ 1 & 0 \end{pmatrix}.$$

It should be noticed that from the view point of Bayesian statistics, the above Hamiltonian (2) corresponds to the logarithm of the posterior distribution. For various such choices of parameters $\{J_{ij}\}, h$ and $\{\tau\}$, one can model the problem of information processing appropriately. For instance,

- For the choice of $J_{ij} = J, h \neq 0$, the above Hamiltonian (2) describes *image restoration*,[26,28,29] which is the problem to estimate the original image from a given set of degraded pixels $\boldsymbol{\tau} = (\tau_1, \cdots, \tau_N)$ by means of Markov random fields. The problem will be dealt with in this manuscript.
- For $J_{ij} \neq 0, h = 0$, (2) corresponds to the logarithm of the posterior of the so-called *Sourlas codes*,[32,33] which is achieved by sending the parity check codes as two-body interactions $\xi_i\xi_j \; \forall(i,j)$ through the binary symmetric channel (BSC).
- For the *Sherrington-Kirkpatrick model*,[23] we may choose J_{ij} obeying the Gaussian with J_0 mean and $\tilde{J}^2$ variance in (2).
- Especially, for the choice of the so-called Hebb rule (Hebbian connections) $J_{ij} = \sum_{\mu=1}^{\alpha N} \xi_i^\mu \xi_j^\mu, h = 0$ in (2), it becomes energy function of the *Hopfield model*[19,20] in which extensive (loading rate $\alpha \neq 0$) /non-extensive (loading rate $\alpha = 0$) number of patterns are embedded. In this paper, we discuss the pattern-recalling process of this model for the case of $\alpha = 0$.

In this section, we should first focus on the simplest case of $J_{ij} = J > 0, h = 0$ ferromagnetic transverse Ising model.[34,35] Let us start our argument from the effective Hamiltonian which is decomposed from (2) by the Suzuki-Trotter formula (1).

$$\beta H = -\sum_{k=1}^{M}\sum_{i=1}^{N}\beta\phi_i(\boldsymbol{\sigma}_k:\sigma_i(k+1))\sigma_i(k)$$

$$= -\frac{\beta J}{MN}\sum_{k,ij}\sigma_i(k)\sigma_j(k) - B\sum_{k,i}\sigma_i(k)\sigma_i(k+1) \quad (3)$$

where k means the Trotter index and M denotes the number of the Trotter slices. We also defined the parameter B as

$$B \equiv \frac{1}{2} \log \coth\left(\frac{\beta\Gamma}{M}\right). \quad (4)$$

A microscopic 'classical' spin state on the k-th Trotter slice is defined by

$$\boldsymbol{\sigma}_k \equiv (\sigma_1(k), \sigma_2(k), \cdots, \sigma_N(k)), \; \sigma_i(k) \in \{+1, -1\}.$$

2.1. *The Glauber dynamics and its transition probability*

In the expression of the effective Hamiltonian (3), $\beta\phi_i(\boldsymbol{\sigma}_k, \sigma_i(k+1))$ is a local field on the cite i in the k-th Trotter slice, which is explicitly given by

$$\beta\phi_i(\boldsymbol{\sigma}_k : \sigma_i(k \pm 1)) = \frac{\beta J}{MN} \sum_j \sigma_j(k) + \frac{B}{2}\{\sigma_i(k-1) + \sigma_i(k+1)\} \quad (5)$$

where parameter B is related to the amplitude of the transverse field (the strength of the quantum-mechanical noise) Γ by equation (4). In the classical limit $\Gamma \to 0$, the parameter B goes to infinity.

Then, the transition probability which specifies the Glauber dynamics of the system is given by

$$w_i(\boldsymbol{\sigma}_k) = \frac{1}{2}[1 - \sigma_i(k)\tanh(\beta\phi_i(\boldsymbol{\sigma}_k : \sigma(k \pm 1)))].$$

More explicitly, $w_i(\boldsymbol{\sigma}_k)$ denotes the probability that an arbitrary classical spin $\sigma_i(k)$ changes its state as $\sigma_i(k) \to -\sigma_i(k)$ within the minimal time unit. Hence, the probability that the spin $\sigma_i(k)$ takes $+1$ is obtained by setting $\sigma_i(k) = -1$ in the above $w_i(\boldsymbol{\sigma}_k)$ and we immediately find $\sigma_i(k) = \sigma_i(k-1) = \sigma_i(k+1)$ with probability 1 in the limit of $B \to \infty$ which implies the classical limit $\Gamma \to 0$.

2.2. *The master equation*

Hence, the probability that a microscopic state including the M-Trotter slices $\{\boldsymbol{\sigma}_k\} \equiv (\boldsymbol{\sigma}_1, \cdots, \boldsymbol{\sigma}_M), \boldsymbol{\sigma}_k \equiv (\sigma_1(k), \cdots, \sigma_N(k))$ obeys the following master equation:

$$\frac{dp_t(\{\boldsymbol{\sigma}_k\})}{dt} = \sum_{k=1}^{M}\sum_{i=1}^{N}[p_t(F_i^{(k)}(\boldsymbol{\sigma}_k))w_i(F_i^{(k)}(\boldsymbol{\sigma}_k)) - p_t(\boldsymbol{\sigma}_k)w_i(\boldsymbol{\sigma}_k)] \quad (6)$$

where $F_i^{(k)}(\cdot)$ denotes a single spin flip operator for neuron i on the Trotter slice k as $\sigma_i(k) \to -\sigma_i(k)$.

2.3. From master equation to deterministic flows

When we pick up a set of magnetizations

$$m_k \equiv \frac{1}{N}\sum_i \sigma_i(k), \quad k = 1, \cdots, M \tag{7}$$

as relevant macroscopic quantities, the joint distribution of the set of the magnetization $\{m_1, \cdots, m_M\}$ at time t is written in terms of the probability for realizations of microscopic states $p_t(\{\boldsymbol{\sigma}_k\})$ at the same time t as

$$P_t(m_1, \cdots, m_M) = \sum_{\{\boldsymbol{\sigma}_k\}} p_t(\{\boldsymbol{\sigma}_k\}) \prod_{k=1}^{M} \delta(m_k - m_k(\boldsymbol{\sigma}_k)) \tag{8}$$

where we defined the sums by

$$\sum_{\{\boldsymbol{\sigma}_k\}}(\cdots) \equiv \sum_{\boldsymbol{\sigma}_1} \cdots \sum_{\boldsymbol{\sigma}_M}(\cdots), \quad \sum_{\boldsymbol{\sigma}_k}(\cdots) \equiv \sum_{\sigma_1(k)=\pm 1} \cdots \sum_{\sigma_N(k)=\pm 1}(\cdots). \tag{9}$$

Taking the derivative of equation (8) with respect to time t and substituting (6) into the result, we have the following differential equations for the joint distribution

$$\frac{dP_t(\{m_k\})}{dt} = \sum_k \frac{\partial}{\partial m_k}\{m_k P_t(\{m_k\})\}$$

$$-\sum_k \frac{\partial}{\partial m_k}\left\{P_t(\{m_k\})\frac{\sum_{\{\boldsymbol{\sigma}_k\}} p_t(\{\boldsymbol{\sigma}_k\}) \tanh[\beta\phi(k)] \prod_{k,i} \delta(m_k - m_k(\boldsymbol{\sigma}_k))}{\sum_{\{\boldsymbol{\sigma}_k\}} p_t(\{\boldsymbol{\sigma}_k\}) \prod_k \delta(m_k - m_k(\boldsymbol{\sigma}_k))}\right\}$$

$$\times \delta(\sigma(k+1) - \sigma_i(k+1))\delta(\sigma(k-1) - \sigma_i(k-1)) \tag{10}$$

where we use the definitions

$$\{m_k\} \equiv (m_1, \cdots, m_M) \tag{11}$$

$$\beta\phi(k) \equiv \frac{\beta}{M} m_k + \frac{B}{2}\sigma(k-1) + \frac{B}{2}\sigma(k+1) \tag{12}$$

as a matter of convenience.

If the local field $\beta\phi(k)$ is independent of the microscopic variable $\{\boldsymbol{\sigma}_k\}$, one can get around the complicated expectation of the quantity $\tanh[\beta\phi(k)]$ over the time-dependent Gibbs measurement which is defined in the so-called 'sub-shell': $\prod_k \delta(m_k - m_k(\boldsymbol{\sigma}_k))$. However, unfortunately, we clearly find from equation (12) that the local field depends on the microscopic state $\{\boldsymbol{\sigma}_k\}$. To overcome the difficulty and to carry out the calculation, we assume that the probability $p_t(\{\boldsymbol{\sigma}_k\})$ of realizations for microscopic states during the dynamics is independent of t, namely,

$$p_t(\{\boldsymbol{\sigma}_k\}) = p(\{\boldsymbol{\sigma}_k\}). \tag{13}$$

Then, our average over the time-dependent Gibbs measurement in the sub-shell is rewritten as

$$\frac{\sum_{\{\boldsymbol{\sigma}_k\}} p_t(\{\boldsymbol{\sigma}_k\}) \tanh[\beta\phi(k)] \prod_{k,i} \delta(m_k - m_k(\boldsymbol{\sigma}_k))}{\sum_{\{\boldsymbol{\sigma}_k\}} p_t(\{\boldsymbol{\sigma}_k\}) \prod_k \delta(m_k - m_k(\boldsymbol{\sigma}_k))}$$
$$\times \delta(\sigma(k+1) - \sigma_i(k+1))\delta(\sigma(k-1) - \sigma_i(k-1))$$
$$\equiv \langle \tanh[\beta\phi(k)] \prod_i \delta(\sigma(k+1) - \sigma_i(k+1))\delta(\sigma(k-1) - \sigma_i(k-1))\rangle_*$$

(14)

where $\langle \cdots \rangle_*$ stands for the average in the sub-shell (constraint) defined by $m_k = m_k(\boldsymbol{\sigma}_k) \ (\forall k)$:

$$\langle \cdots \rangle_* \equiv \frac{\sum_{\{\boldsymbol{\sigma}_k\}} p(\{\boldsymbol{\sigma}_k\})(\cdots) \prod_k \delta(m_k - m_k(\boldsymbol{\sigma}_k))}{\sum_{\{\boldsymbol{\sigma}_k\}} p(\{\boldsymbol{\sigma}_k\}) \prod_k \delta(m_k - m_k(\boldsymbol{\sigma}_k))} \quad (15)$$

If we notice that the above Gibbs measurement (15) in the sub-shell is rewritten as

$$\sum_{\{\boldsymbol{\sigma}_k\}} p(\{\boldsymbol{\sigma}_k\}) \prod_k \delta(m_k - m_k(\boldsymbol{\sigma}_k)) = \text{tr}_{\{\sigma\}} \exp\left[\beta \sum_{l=1}^{M} \phi(l)\sigma(l)\right] \quad (16)$$

$(\text{tr}_{\{\sigma\}}(\cdots) \equiv \prod_k \sum_{\boldsymbol{\sigma}_k}(\cdots))$, and the quantity

$$\tanh[\beta\phi(k)] = \frac{\sum_{\sigma(k)=\pm 1} \sigma(k) \exp[\beta\phi(k)\sigma(k)]}{\sum_{\sigma(k)=\pm 1} \exp[\beta\phi(k)\sigma(k)]} \quad (17)$$

is independent of $\sigma(k)$, the average appearing in (14) leads to

$$\langle \tanh[\beta\phi(k)] \prod_i \delta(\sigma(k+1) - \sigma_i(k+1))\delta(\sigma(k-1) - \sigma_i(k-1))\rangle_*$$
$$= \frac{\text{tr}_{\{\sigma\}}\{\frac{1}{M}\sum_{l=1}^{M}\sigma(l)\} \exp[\beta\phi(k)\sigma(k)]}{\text{tr}_{\{\sigma\}} \exp[\beta\phi(k)\sigma(k)]} \equiv \langle \sigma \rangle_{path} \quad (18)$$

in the limit of $M \to \infty$. This is nothing but a path integral for the 'effective single spin' problem in which the spin $\sigma(k)$ updates its state along the imaginary-time axis: $\text{tr}_{\{\sigma\}}(\cdots) \equiv \sum_{\sigma(1)=\pm 1} \cdots \sum_{\sigma(M)=\pm 1}(\cdots)$ with weights $\exp[\beta\phi(k)\sigma(k)], \ (k = 1, \cdots, M)$.

Then, the differential equation (10) leads to

$$\frac{dP_t(\{m_k\})}{dt} = \sum_k \frac{\partial}{\partial m_k}\{m_k P_t(\{m_k\})\} - \sum_k \frac{\partial}{\partial m_k}\left\{P_t(\{m_k\})\langle \sigma \rangle_{path}\right\}.$$

(19)

In order to derive the compact form of the differential equations with respect to the overlaps, we substitute

$$P_t(\{m_k\}) = \prod_{k=1}^{M} \delta(m_k - m_k(t))$$

into the above (19) and multiplying m_l by both sides of the equation and carrying out the integral with respect to $dm_1 \cdots dm_M$ by part, we have for $l = 1, \cdots, M$ as

$$\frac{dm_l}{dt} = -m_l + \langle \sigma \rangle_{path}. \tag{20}$$

In the next subsection, we carry out the average $\langle \cdots \rangle_{path}$ as a path integral explicitly under the so-called static approximation.

2.3.1. *The static approximation*

In order to obtain the final form of the deterministic flow, we assume that macroscopic quantities such as the magnetization m_k are independent of the Trotter slices k during the dynamics. Namely, we must use the so-called *static approximation*:

$$m_k = m \; (\forall k). \tag{21}$$

Under the static approximation, let us use the following inverse process of the Suzuki-Trotter decomposition (1):

$$\lim_{M \to \infty} Z_M = \text{tr} \exp\left[\beta(Jm\,\sigma_z + \Gamma\sigma_x)\right] \tag{22}$$

$$Z_M \equiv \text{tr}_{\{\sigma\}} \exp\left[\frac{\beta Jm}{M}\sum_k \sigma(k) + B\sum_k \sigma(k)\sigma(k+1)\right] \tag{23}$$

Then, one can calculate the path integral immediately as

$$\langle \sigma \rangle_{path} = \frac{Jm}{\sqrt{(Jm)^2 + \Gamma^2}} \tanh \beta\sqrt{(Jm)^2 + \Gamma^2}. \tag{24}$$

Inserting this result into (20), we obtain

$$\frac{dm}{dt} = -m + \frac{Jm}{\sqrt{(Jm)^2 + \Gamma^2}} \tanh \beta\sqrt{(Jm)^2 + \Gamma^2}. \tag{25}$$

2.3.2. Classical limit

By taking the limit $\Gamma \to 0$ in the equation (25), we immediately obtain the spontaneous magnetization flow for the dynamical Ising model:

$$\frac{dm}{dt} = -m + \tanh(\beta J m). \qquad (26)$$

Near the critical point, it behaves as $dm/dt = -(1 - \beta J)m - (\beta J m)^3/3 + \mathcal{O}(m^5)$. From this equation, we easily find that spontaneous magnetization shows well-known time-dependent behaviour

$$m(t) \simeq m(0)\, e^{-t/(1-\beta J)^{-1}}$$

around the critical point $\beta \simeq \beta_c = J^{-1}$, and at the critical point, the relaxation time diverges as $(1 - \beta_c J)^{-1}$ resulting in $m(t) \simeq t^{-1/2}$ (critical slowing down with dynamical exponent $\nu' = 1/2$).

2.4. Dynamics and steady state

It is easy for us to confirm that the steady state indicated by $dm/dt = 0$ in (25) is nothing but the equilibrium state described by the equation of state:

$$m = \frac{Jm}{\sqrt{(Jm)^2 + \Gamma^2}} \tanh \beta \sqrt{(Jm)^2 + \Gamma^2}.$$

In Fig. 1(left), we plot the typical behaviour of zero-temperature dynamics (equation (25) with $\beta = \infty$) far from the critical point $\Gamma_c = J = 1$ of

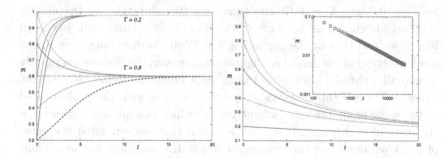

Fig. 1. Typical behaviour of zero-temperature dynamics described by (25) with $\beta = \infty$ far from the critical point $\Gamma_c = J = 1$ of quantum phase transition (left). The right panel denotes the zero-temperature dynamics at the critical point. The inset shows the log-log plot of $m(t)$ indicating that the dynamical exponent in the critical slowing down is $\nu' = 1/2$.

quantum phase transition. We easily find that the dynamics exponentially converges to the steady state. The right panel denotes the zero-temperature dynamics at the critical point. The inset shows the log-log plot of $m(t)$ indicating that the dynamical exponent in the critical slowing down is $\nu' = 1/2$. This fact is directly confirmed from equation (25) with $\beta = \infty$ near the critical point

$$\frac{dm}{dt} \simeq -(1 - J\Gamma^{-1})m - (Jm)^3/2\Gamma^3 + \mathcal{O}(m^5), \qquad (27)$$

namely, $m(t)$ behaves around the critical point $\Gamma \simeq \Gamma_c = J$ as

$$m(t) = m(0)\,e^{-t/(1-J\Gamma^{-1})^{-1}}.$$

At the critical point, the relaxation time diverges as $\tau_\Gamma \equiv (1 - J\Gamma_c^{-1})^{-1}$ resulting in the critical slowing down as $m(t) \simeq e^{-t/\tau_\Gamma} \to m(t) \simeq t^{-\nu'}$, $\nu' = 1/2$. Of course, the exponent is the same as that of the 'mean-field model' universality class.

2.5. On the validity of static approximation

To confirm the validity of the static approximation, we carry out computer simulation for finite size system having $N = 400$ spins. We observe the time evolving process of the histogram $P(m_k)$ which is calculated from the $M = N = 400$ copies of the Trotter slices.

We show the result in Fig. 2. In this simulation, we chose the initial configuration in each Trotter slice randomly (we set each spin variable $\sigma_i(k)$ to $+1$ with a fixed probability p) and choose the inverse temperature $\beta = 2$ for $\Gamma = 0.5$ and $\Gamma = 0.6$. The time unit (the duration) of the update of $P(m_k)$ is chosen as 1 Monte Carlo step (MCS). From both panels in Fig. 2, we find that at the beginning, the $P(m_k)$ is distributed due to the random set-up of the initial configuration, however, the fluctuation rapidly (eventually) shrinks leading up to the delta function around MCS ~ 100. After that, the $P(m_k)$ evolves as a delta function with the peak located at the value of spontaneous magnetization which is explicitly indicated in the inset of each panel. It should be noted that we evaluated the value of order parameter at the time point in the Runge-Kutta method. Thus, the duration between the points to be evaluated is not the MCS but the Runge-Kutta step.

Of course, some statistical errors for the finite system should be taken into account, however, the limited result here seems to support the validity of the static approximation even in the dynamical process. The validity of

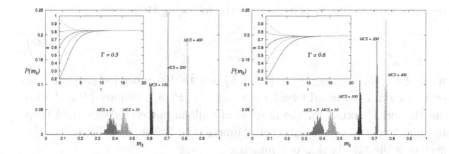

Fig. 2. Time evolution of the distribution $P(m_k)$ calculated for finite size system with $N = M = 400$. We choose the inverse temperature $\beta = 2$ for $\Gamma = 0.5$ (left) and $\Gamma = 0.6$ (right). The inset in each panel denotes the deterministic flows of spontaneous magnetization calculated by (25) for corresponding parameter sets.

the static approximation was recently argued by Takahashi and Matsuda[48] from the different perspective.

3. The quantum Hopfield model

From this section, we shall apply our formulation to several problems being outside of physics. In this section, we first briefly explain the basics of the conventional Hopfield model as a possible application of our formula. Then, we shall divide the model into two classes, namely, the Hopfield model put in thermal noises (the model is referred to as *classical systems*) and the same model in the quantum-mechanical noise (the model is referred to as *quantum systems*). In following, we define each of the models explicitly.

3.1. The classical system

Let us consider the network having N-neurons. Each neuron S_i takes two states, namely,

$$S_i = \begin{cases} +1 \ (\textit{firing}) \\ -1 \ (\textit{stationary}) \end{cases} \quad (28)$$

Neuronal states are given by the set of variables S_i, that is,

$$\boldsymbol{S} = (S_1, \cdots, S_N),\ S_i \in \{+1, -1\}.$$

Each neuron is located on a complete graph, namely, graph topology of the network is 'fully-connected'. The synaptic connection between arbitrary two neurons, say, S_i and S_j is defined by the following Hebb rule:

$$J_{ij} = \frac{1}{N} \sum_{\mu,\nu} \xi_i^\mu A_{\mu\nu} \xi_j^\nu \qquad (29)$$

where $\boldsymbol{\xi}^\mu = (\xi_1, \cdots, \xi_N), \xi_i^\mu \in \{+1, -1\}$ denote the embedded patterns and each of them is specified by a label $\mu = 1, \cdots, P$. $A_{\mu\nu}$ denotes $(P \times P)$-size matrix and P stands for the number of built-in patterns. We should keep in mind that there exists an energy function (a Lyapunov function) in the system if the matrix $A_{\mu\nu}$ is symmetric.

Then, the output of the neuron i, that is, S_i is determined by the sign of the local field h_i as

$$h_i = \sum_{\mu,\nu=1}^{p} \xi_i^\mu A_{\mu\nu} m^\nu + \frac{1}{N} \sum_{a,b=p+1}^{P} \xi_i^a A_{ab} \sum_j \xi_j^b S_j \qquad (30)$$

where $A_{\mu\nu}$ and A_{ab} are elements of $p \times p, (P-p) \times (P-p)$-size matrices, respectively. We also defined the overlap (the direction cosine) between the state of neurons $\boldsymbol{S}$ and one of the built-in patterns $\boldsymbol{\xi}^\nu$ by

$$m^\nu \equiv \frac{1}{N}(\boldsymbol{S} \cdot \boldsymbol{\xi}^\nu) = \frac{1}{N} \sum_i \xi_i^\nu S_i. \qquad (31)$$

Here we should notice that the Hamiltonian of the system is given by $-\sum_i h_i S_i$. The first term appearing in the left hand side of equation (30) is a contribution from $p \sim \mathcal{O}(1)$ what we call 'condensed patterns', whereas the second term stands for the so-called 'cross-talk noise'. In this paper, we shall concentrate ourselves to the case in which the second term is negligibly small in comparison with the first term, namely, the case of $P = p \sim \mathcal{O}(1)$. In this sense, we can say that the network is 'far from its saturation'.

3.2. The quantum system

In order to extend the classical system to the quantum-mechanical variant, we shall rewrite the local field h_i as follows.

$$\phi_i = \sum_{\mu,\nu=1}^{p} \xi_i^\mu A_{\mu\nu} \left(\frac{1}{N} \sum_i \xi_i^\nu \sigma_i^z \right) \qquad (32)$$

where σ_i^z ($i = 1, \cdots, N$) stands for the z-component of the Pauli matrix. Thus, the Hamiltonian

$$\boldsymbol{H}_0 \equiv -\sum_i \phi_i \sigma_i^z$$

is a diagonalized ($2^N \times 2^N$)-size huge matrix and the lowest eigenvalue is identical to the ground state of the classical Hamiltonian $-\sum_i \phi_i S_i$ (S_i is an eigenvalue of the matrix σ_i^z).

Then, we introduce quantum-mechanical noise into the Hopfield neural network by adding the transverse field to the Hamiltonian as follows.

$$H = H_0 - \Gamma \sum_{i=1}^{N} \sigma_i^x \qquad (33)$$

where σ_i^x is the x-component of the Pauli matrix and transitions between eigenvectors of the classical Hamiltonian H_0 are induced due to the off-diagonal elements of the matrix H for $\Gamma \neq 0$. In this paper, we mainly consider the system described by (33).

3.2.1. *The master equation*

As we saw in the pure ferromagnetic Ising model, by making use of the Suzuki-Trotter decomposition (1), we obtain the local field for the neuron i located on the k-th Trotter slice as follows.

$$\beta \phi_i(\sigma_k : \sigma_i(k \pm 1)) = \frac{\beta}{M} \sum_{\mu,\nu} \xi_i^\nu A_{\mu\nu} \left\{ \frac{1}{N} \sum_j \xi_j^\nu \sigma_j(k) \right\}$$
$$+ \frac{B}{2} \{\sigma_i(k-1) + \sigma_i(k+1)\} \qquad (34)$$

Then, we pick up the overlap between neuronal state σ_k and one of the built-in patterns ξ^ν, namely,

$$m_k \equiv \frac{1}{N}(\sigma_k \cdot \xi^\nu) = \frac{1}{N} \sum_i \xi_i^\nu \sigma_i(k) \qquad (35)$$

as a relevant macroscopic quantity, the joint distribution of the set of the overlaps:

$$\{m_k^\nu\} \equiv (m_1^\nu, \cdots, m_M^\nu)$$

at time t is written in terms of the probability for realizations of microscopic states $p_t(\{\sigma_k\})$ at the same time t as

$$P_t(\{m_k^\nu\}) = \sum_{\{\sigma_k\}} p_t(\{\sigma_k\}) \prod_{k=1}^{M} \delta(m_k^\nu - m_k^\nu(\sigma_k)) \qquad (36)$$

Using the same way as in the previous section (under the approximations: $p_t(\{\boldsymbol{\sigma}_k\}) = p(\{\boldsymbol{\sigma}_k\})$ and $m_k^\nu = m^\nu \ \forall k$), we have

$$\frac{dm^\nu}{dt} = -m^\nu + \mathbb{E}_{\boldsymbol{\xi}}\left[\frac{\xi^\nu \sum_{\mu\nu} \xi^\mu A_{\mu\nu} m^\nu}{\sqrt{(\sum_{\mu\nu} \xi^\mu A_{\mu\nu} m^\nu)^2 + \Gamma^2}} \tanh\beta\sqrt{\left(\sum_{\mu\nu} \xi^\mu A_{\mu\nu} m^\nu\right)^2 + \Gamma^2}\right] \quad (37)$$

where we should bear in mind that the empirical distribution $D[\xi^\nu]$:

$$D[\xi^\nu] \equiv \frac{1}{N}\sum_i \delta(\xi^\nu - \xi_i^\nu) \quad (38)$$

was replaced by the built-in pattern distribution $\mathcal{P}(\xi^\nu)$ as

$$\lim_{N\to\infty} \frac{1}{N}\sum_i \delta(\xi_i^\nu - \xi^\nu) = \mathcal{P}(\xi^\nu) \quad (39)$$

in the limit of $N \to \infty$ and the average is now carried out explicitly as

$$\prod_{\nu=1}^p \int D[\xi^\nu] d\xi^\nu(\cdots) = \prod_{\nu=1}^p \int \mathcal{P}(\xi^\nu) d\xi^\nu(\cdots) \equiv \mathbb{E}_{\boldsymbol{\xi}}[\cdots], \ p = \mathcal{O}(1). \quad (40)$$

Equation (37) is a general solution for the pattern-recalling problem.

3.2.2. *The classical and zero-temperature limits*

It is easy for us to take the classical limit $\Gamma \to 0$ in the result (37). Actually, we have immediately

$$\frac{dm^\nu}{dt} = -m^\nu + \mathbb{E}_{\boldsymbol{\xi}}\left[\xi^\nu \tanh\left(\beta \sum_{\mu\nu} \xi^\mu A_{\mu\nu} m^\nu\right)\right]. \quad (41)$$

The above equation is identical to the result by Coolen and Ruijgrok[18] who considered the retrieval process of the conventional Hopfield model under thermal noise.

We can also take the zero-temperature limit $\beta \to \infty$ in (37) as

$$\frac{dm^\nu}{dt} = -m^\nu + \mathbb{E}_{\boldsymbol{\xi}}\left[\frac{\xi^\nu \sum_{\mu\nu} \xi^\mu A_{\mu\nu} m^\nu}{\sqrt{(\sum_{\mu\nu} \xi^\mu A_{\mu\nu} m^\nu)^2 + \Gamma^2}}\right]. \quad (42)$$

Thus, the equation (37) including the above two limiting cases is our general solution for the neurodynamics of the quantum Hopfield model in which $\mathcal{O}(1)$ patterns are embedded. Thus, we can discuss any kind of situations for such pattern-recalling processes and the solution is always derived from equation (37) explicitly.

3.3. Limit cycle solution for asymmetric connections

In this section, we shall discuss a specific case of the general solution (37). Namely, we investigate the pattern-recalling processes of the quantum Hopfield model with asymmetric connections $A \equiv \{A_{\mu\nu}\}$.

3.3.1. Result for just only two embedded patterns

Let us consider the case in which just only two patterns are embedded via the following matrix:

$$A = \begin{pmatrix} 1 & -1 \\ 1 & 1 \end{pmatrix} \qquad (43)$$

Then, from the general solution (37), the differential equations with respect to the two overlaps m_1 and m_2 are written as

$$\frac{dm_1}{dt} = -m_1 + \frac{m_1}{\sqrt{(2m_1)^2 + \Gamma^2}} - \frac{m_2}{\sqrt{(2m_2)^2 + \Gamma^2}}$$

$$\frac{dm_2}{dt} = -m_2 + \frac{m_1}{\sqrt{(2m_1)^2 + \Gamma^2}} + \frac{m_2}{\sqrt{(2m_2)^2 + \Gamma^2}}.$$

In Fig. 3, we show the time evolutions of the overlaps m_1 and m_2 for the case of the amplitude $\Gamma = 0.01$. From this figure, we clearly find that the

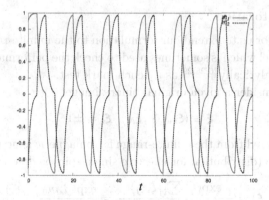

Fig. 3. Time evolutions of m_1 and m_2 for the case of $\Gamma = 0.01$.

neuronal state evolves as $A \to B \to \overline{A} \to \overline{B} \to A \to B \to \cdots$ ($\overline{A}, \overline{B}$ denote the 'mirror images' of A and B, respectively), namely, the network behaves as a limit cycle.

To compare the effects of thermal and quantum noises on the pattern-recalling processes, we plot the trajectories m_1-m_2 for $(T \equiv \beta^{-1}, \Gamma) = (0, 0.01), (0.01, 0)$ (left panel), $(T, \Gamma) = (0, 0.8), (0.8, 0)$ (right panel) in Fig. 4. From these panels, we find that the limit cycles are getting col-

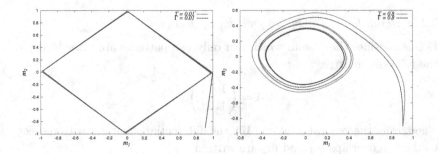

Fig. 4. Typical trajectories m_1-m_2 for $(T, \Gamma) = (0, 0.01), (0.01, 0)$ (left panel), $(T, \Gamma) = (0, 0.8), (0.8, 0)$ (right panel).

lapsed as the strength of the noise level is increasing for both thermal and quantum-mechanical noises, and eventually the trajectories shrink to the origin $(m_1, m_2) = (0, 0)$ in the limit of $T, \Gamma \to \infty$.

4. Image restoration

It is also easy for us to extend our formulation to the infinite-range random field Ising model which is sometimes used to check the performance of image restoration analytically[26,28,29] as a bench mark test.

Here we consider a given original binary image:

$$\boldsymbol{\xi} \equiv (\xi_1, \cdots, \xi_N), \ \xi_i = \pm 1$$

which is generated from the infinite-range ferromagnetic Ising model whose Gibbs measure (distribution for effective single spin) is described by

$$P(\boldsymbol{\xi}) = \frac{\exp(\frac{\beta_s}{N} \sum_{ij} \xi_i \xi_j)}{\sum_{\boldsymbol{\xi}} \exp(\frac{\beta_s}{N} \sum_{ij} \xi_i \xi_j)} = \frac{\exp(\beta_s m_0 \sum_i \xi_i)}{\{2 \cosh(\beta_s m_0)\}^N} \quad (44)$$

where m_0 denotes spontaneous magnetization at temperature β_s^{-1} and specifies the original image $\boldsymbol{\xi}$ macroscopically. m_0 is given by a solution of the equation of state (for a pure ferromagnetic Ising model):

$$m_0 = \tanh(\beta_s m_0) \quad (45)$$

A snapshot $\boldsymbol{\xi}$ from the distribution (44) is degraded by additive white Gaussian noise (AWGN) with mean $a_0 \xi_i$ and variance a^2, namely, each pixel in the degraded image $\boldsymbol{\tau} = (\tau_1, \cdots, \tau_N)$ is obtained by

$$\tau_i = a_0 \xi_i + a x_i, \quad x_i \sim \mathcal{N}(0,1).$$

More explicitly, the conditional probability for describing the degrading process $P(\boldsymbol{\tau}|\boldsymbol{\xi}) = \prod_i P(\tau_i|\xi_i)$ is given by

$$P(\boldsymbol{\xi}|\boldsymbol{\tau}) = \frac{\exp\left[-\frac{1}{2a^2}\sum_i(\tau_i - a_0\xi_i)^2\right]}{(\sqrt{2\pi}a)^N} = \frac{e^{-\frac{1}{2a^2}\sum_i(\tau_i^2+a_0^2)}\exp\left(\frac{a_0}{a^2}\sum_i \tau_i \xi_i\right)}{\int_{-\infty}^{\infty}\prod_i d\tau_i\, e^{-\frac{1}{2a^2}\sum_i(\tau_i^2+a_0^2)}}. \tag{46}$$

From the Bayesian inference view point, one can assume that the posterior $P(\boldsymbol{\sigma}|\boldsymbol{\tau})$ (here we define $\boldsymbol{\sigma}$ as an estimate of the original image $\boldsymbol{\xi}$) might be proportional to the logarithm of the effective Hamiltonian:

$$H = -\sum_i \sigma_i \phi_i(\boldsymbol{\sigma}) = -\frac{J}{N}\sum_{ij}\sigma_i \sigma_j + h \sum_i \tau_i \sigma_i \tag{47}$$

with

$$\phi_i(\boldsymbol{\sigma}) \equiv \frac{J}{N}\sum_j \sigma_j + h \tau_i \sigma_i. \tag{48}$$

where J and h are basically unknown parameters (what we call, *hyperparameters*) and the best possible choice is apparently to set the J to the corresponding parameter β_s in (44) as $J = \beta_s$, and to set h to the 'signal-to-noise ratio' in the degrading process as $h = a_0/a^2$.

To introduce 'quantum fluctuation' into the system, we shall rewrite the above Hamiltonian (47) by the quantum-mechanical variant (2) with $J_{ij} = J\,\forall(i,j)$ and $h \neq 0$. The first and the second terms appearing in the right hand side of (2) correspond to the prior distribution and the likelihood function, respectively. Whereas, the third term is introduced to utilize quantum fluctuation to construct the Bayesian estimate for each pixel ('majority-vote decision' on each pixel), namely, $\mathrm{sgn}(\langle \sigma_i^z \rangle)$.

In this section, we attempt to describe the recovering process of original image through the deterministic flows of several relevant order-parameters and image restoration measure, namely, the overlap function

$$M \equiv \frac{1}{N}\sum_i \xi_i\, \mathrm{sgn}(\langle \sigma_i^z \rangle). \tag{49}$$

For the above set-up of the problem, the local field on the site i in the k-th Trotter slice now leads to

$$\beta\phi_i(\boldsymbol{\sigma}_k:\sigma_i(k\pm 1),\boldsymbol{\tau}) = \frac{\beta J}{NM}\sum_j \sigma_j(k) + \frac{\beta h}{M}\tau_i\sigma_i(k)$$
$$+ \frac{B}{2}\{\sigma_i(k-1)+\sigma_i(k+1)\}. \quad (50)$$

As relevant order parameters, we choose the overlap between the degraded image $\boldsymbol{\tau}$ and an estimate of the original image $\boldsymbol{\sigma}_k$ as

$$\mu_k = \frac{1}{N}\sum_i \tau_i\sigma_i(k) \quad (51)$$

and magnetization m_k. Then, we derive the differential equation with respect to

$$P_t(\{m_k\},\{\mu_k\}) = \sum_{\{\boldsymbol{\sigma}_k\}} p_t(\{\boldsymbol{\sigma}_k\})\prod_{k=1}^M \delta(m_k - m_k(\boldsymbol{\sigma}_k))\delta(\mu_k - \mu_k(\boldsymbol{\sigma}_k)) \quad (52)$$

as follows.

$$\frac{dP_t(\{m_k\},\{\mu_k\})}{dt}$$
$$= \sum_k \left(\frac{\partial}{\partial m_k}\{m_k P_t(\{m_k\},\{\mu_k\})\} + \frac{\partial}{\partial \mu_k}\{\mu_k P_t(\{m_k\},\{\mu_k\})\}\right)$$
$$- \sum_k \frac{\partial}{\partial m_k}\left\{P_t(\{m_k\},\{\mu_k\})\frac{1}{N}\sum_i \tanh[\beta\phi_i(\boldsymbol{\sigma}_k:\sigma_i(k\pm 1),\boldsymbol{\tau})]\right\}$$
$$- \sum_k \frac{\partial}{\partial \mu_k}\left\{P_t(\{m_k\},\{\mu_k\})\frac{1}{N}\sum_i \tau_i\tanh[\beta\phi_i(\boldsymbol{\sigma}_k:\sigma_i(k\pm 1),\boldsymbol{\tau})]\right\}$$
$$(53)$$

where we defined

$$\{m_k\} = (m_1,\cdots,m_M),\ \{\mu_k\} = (\mu_1,\cdots,\mu_M).$$

By assuming the self-averaging properties on the following physical quantities over both all possible paths in the imaginary-time axis and input data; original images and degrading processes (a particular realization of the quantity is identical to the average value and its deviation from the average eventually vanishes in the limit $N \to \infty$), we have

$$\lim_{N \to \infty} \frac{1}{N} \sum_i \tanh[\beta \phi_i(\boldsymbol{\sigma}_k : \sigma_i(k \pm 1), \boldsymbol{\tau})] = \mathbb{E}_{\boldsymbol{\tau}}[\langle \sigma(k) \rangle_{*path}] \quad (54)$$

$$\lim_{N \to \infty} \frac{1}{N} \sum_i \tau_i \tanh[\beta \phi_i(\boldsymbol{\sigma}_k : \sigma_i(k \pm 1), \boldsymbol{\tau})] = \mathbb{E}_{\boldsymbol{\tau}}[\tau \langle \sigma(k) \rangle_{*path}] \quad (55)$$

where we defined the two different kinds of the averages by

$$\langle \cdots \rangle_{*path} \equiv \lim_{M \to \infty} \frac{\mathrm{tr}_{\{\sigma\}}(\cdots) \exp[\frac{\beta}{M} \sum_{l=1}^{M} \mathcal{C}\sigma(l) + B \sum_{l=1}^{M} \sigma(l)\sigma(l+1)]}{\mathrm{tr}_{\{\sigma\}} \exp[\frac{\beta}{M} \sum_{l=1}^{M} \mathcal{C}\sigma(l) + B \sum_{l=1}^{M} \sigma(l)\sigma(l+1)]}$$

$$\mathbb{E}[\cdots]_\tau \equiv \frac{\sum_\xi e^{\beta_s m_0 \xi}}{2 \cosh(\beta_s m_0)} \int_{-\infty}^{\infty} (\cdots) \exp\left[-\frac{(\tau^2 + a_0^2)}{2a^2}\right] \exp\left(\frac{a_0}{a^2}\tau\xi\right). \quad (56)$$

with $\mathcal{C} \equiv Jm + h\tau$. Under the static approximation, we obtain

$$\frac{dP_t(m,\mu)}{dt} = \frac{\partial}{\partial m}\{mP_t(m,\mu)\} + \frac{\partial}{\partial \mu}\{\mu P_t(m,\mu)\}$$

$$- \frac{\partial}{\partial m}\left\{\frac{P_t(m,\mu) \sum_\xi e^{\beta_s m_0 \xi}}{2\cosh(\beta_s m_0)} \int_{-\infty}^{\infty} Dx \frac{\Xi_m^{(a,a_0)}(\xi,x) \tanh \beta \Phi_{m,\Gamma}^{(a,a_0)}(\xi,x)}{\Phi_{m,\Gamma}^{(a,a_0)}(\xi,x)}\right\}$$

$$- \frac{\partial}{\partial \mu}\left\{\frac{P_t(m,\mu) \sum_\xi e^{\beta_s m_0 \xi}}{2\cosh(\beta_s m_0)} \int_{-\infty}^{\infty} Dx \frac{\mathcal{A}\Xi_m^{(a,a_0)}(\xi,x) \tanh \beta \Phi_{m,\Gamma}^{(a,a_0)}(\xi,x)}{\Phi_{m,\Gamma}^{(a,a_0)}(\xi,x)}\right\}$$

$$(57)$$

with

$$\Xi_m^{(a,a_0)}(\xi,x) \equiv Jm + ha_0\xi + hax, \quad (58)$$

$$\Phi_{m,\Gamma}^{(a,a_0)}(\xi,x) \equiv \sqrt{\{\Xi_m^{(a,a_0)}(\xi,x)\}^2 + \Gamma^2}, \quad \mathcal{A} \equiv a_0\xi + ax \quad (59)$$

and $Dx \equiv dx\, e^{-x^2/2}/\sqrt{2\pi}$. Using the same way as the pure ferromagnetic Ising system, we finally obtain the deterministic flow equations of the order-parameters m and μ as follows.

$$\frac{dm}{dt} = -m + \frac{\sum_\xi e^{\beta_s m_0 \xi}}{2\cosh(\beta_s m_0)} \int_{-\infty}^{\infty} Dx \frac{\Xi_m^{(a,a_0)}(\xi,x) \tanh \beta \Phi_{m,\Gamma}^{(a,a_0)}(\xi,x)}{\Phi_{m,\Gamma}^{(a,a_0)}(\xi,x)} \quad (60)$$

$$\frac{d\mu}{dt} = -\mu + \frac{\sum_\xi e^{\beta_s m_0 \xi}}{2\cosh(\beta_s m_0)} \int_{-\infty}^{\infty} Dx \frac{\mathcal{A}\Xi_m^{(a,a_0)}(\xi,x) \tanh \beta \Phi_{m,\Gamma}^{(a,a_0)}(\xi,x)}{\Phi_{m,\Gamma}^{(a,a_0)}(\xi,x)}.$$

$$(61)$$

For the solution of the above deterministic flows (m, μ) at time t, the overlap between the original image and degraded image is measured by

$$\mathcal{M}(m,\mu) \equiv \frac{1}{N} \sum_i \mathbb{E}_\tau [\xi_i \, \text{sgn}(\langle \sigma_i(k) \rangle_{path})]$$

$$= \frac{\sum_\xi e^{\beta_s m_0 \xi}}{2 \cosh(\beta_s m_0)} \int_{-\infty}^{\infty} Dx \, \text{sgn}[\hat{m} + \mathcal{A}\hat{\mu}] \quad (62)$$

where $(\hat{m}, \hat{\mu})$ is a solution of the following coupled equations

$$m = \frac{\sum_\xi e^{\beta_s m_0 \xi}}{2 \cosh(\beta_s m_0)} \int_{-\infty}^{\infty} Dx \, \tanh[\hat{m} + \mathcal{A}\hat{\mu}] \quad (63)$$

$$\mu = \frac{\sum_\xi e^{\beta_s m_0 \xi}}{2 \cosh(\beta_s m_0)} \int_{-\infty}^{\infty} Dx \, [\hat{m} + \mathcal{A}\hat{\mu}] \tanh[\hat{m} + \mathcal{A}\hat{\mu}] \quad (64)$$

for a given point on the trajectory (m, μ) at time t. To obtain the overlap function (62) and (63)(64), we used the concept of dynamical replica theory (the DRT),[21,22] namely, 'equipartitioning' and 'self-averaging' of the $\mathcal{M}$ during the evolution in time. The detail is available in the reference.[1]

We solve the equations (60)(61) with (62)-(64) numerically and show the results in Fig. 5. We choose the set of the parameters for the original image as $\beta_s^{-1} = 0.9$ ($m_0 = 0.523$) and $a_0 = a = 1$ which means that the corresponding optimal hyper-parameters are $h = a_0/a^2 = 1$ and $J^{-1} = \beta^{-1}$. We consider the zero-temperature restoration dynamics in which the fluctuation to make the Bayesian estimate is only quantum-mechanical one (it is controlled by the amplitude of quantum-mechanical tunneling Γ). In the left panel, the deterministic trajectories in the space (m, μ) are plotted for $\Gamma = 0.6$. The state of system in which the image restoration is successfully achieved is in ferromagnetic phase. Thus, order-parameters m and μ converge to the fixed point exponentially (there is no critical slowing down in this model system). In the right panel, we show the time evolution of the image restoration measure $\mathcal{M}$ for several values of Γ. From this panel, we find that some 'non-monotonic' behaviour is observed at the initial stage of the dynamics when we fail to set the amplitude to its optimal value ($\Gamma \sim 0.6$). Similar behaviour was reported in the Bayesian image restoration via thermal (classical) fluctuation.[38]

In the Bayesian framework, it is desired for us to obtain the estimate of each pixel and hyper-parameters simultaneously. In such case, we use the so-called EM algorithm based on the maximization of marginal likelihood criteria.[39] The quantum-mechanical extension and the formulation

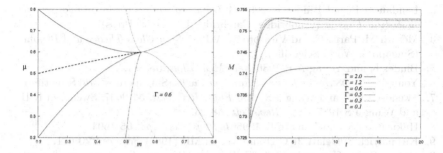

Fig. 5. Trajectories in the phase space (m, μ) are plotted for $\Gamma = 0.6$ (left). The parameters are chosen as $\beta_s^{-1} = 0.9$ ($m_0 = 0.523$), $a_0 = a = 1$ and the corresponding hyper-parameters as $h = a_0/a^2 = 1$ and $J^{-1} = \beta^{-1}$. The right panel shows the evolution of image restoration measure, overlap function $\mathcal{M}$ in time for several values of the amplitude of quantum-mechanical tunneling Γ.

presented here is applicable to the simultaneous estimation for both micro and macro parameters.

5. Concluding remarks

In this paper, for a simplest quantum spin systems, we showed a formulation to describe the macroscopically deterministic flows of order parameters from the master equation whose transition probability is given by the Glauber-type. Under the static approximation, differential equations with respect to macroscopic order parameters were explicitly obtained from the master equation describing the microscopic-law. In the steady state, we found that the equations are identical to the saddle point equations for the equilibrium state of the same system. We also checked the validity of the static approximation by computer simulations and found that the result supports the validity of the approximation. The formula was applied to several problems in the research fields of information science and brain science.

Acknowledgement

The author thanks organizers of *Symposium on Interface between Quantum Information and Statistical Physics*, in particular, Professor Mikio Nakahara and Dr Shu Tanaka.

References

1. Inoue J 2010 *Journal of Physics: Conference Series* **233** 012010.
2. Inoue J 2011 *Journal of Physics: Conference Series* **297** 012012.

3. Landau D P and Binder K 2000 *A Guide to Monte Carlo Simulations in Statistical Physics* (Cambridge: Cambridge University Press).
4. Mézard M, Parisi G and Virasoro M A 1987 *Spin Glass Theory and Beyond* (Singapore: World Scientific).
5. Binder K and Young A P 1986 *Rev. Mod. Phys.* **58** 801.
6. Young A P 1998 *Spin Glass and Random Fields* (Singapore: World Scientific).
7. Swendsen R H and Wang J S 1986 *Phys. Rev. Lett.* **57** 2607, Swendsen R H and Wang J S 1987 *Phys. Rev. Lett.* **58** 86.
8. Hukushima K and Nemoto K 1996 *J. Phys. Soc. Jpn.* **65** 1604.
9. Kirkpatrick S, Gelatt Jr C D and Vecchi M P (1983) *Science* **220** 671.
10. Geman S and Geman D 1984 *IEEE Trans. Pattern. Anal. and Mach. Intel.* **11** 721.
11. Kadowaki T and Nishimori H 1998 *Physical Review E* **58** 5355.
12. Farhi E, Goldstone J, Gutmann S, Lapan J, Lundgren A and Preda P 2001 *Science* **292** 472.
13. Morita S and Nishimori H 2006 *J. Phys. A* **39** 13903.
14. Suzuki S and Okada M 2005 *J. Phys. Soc. Jpn.* **74** 1649.
15. Santoro G E and Tosatti 2006 *J. Phys. A* **41** 209801.
16. Suzuki M 1976 *Prog. Theor. Phys.* **56** 1454.
17. de Oriveira M J and Chiappin J R N 1997 *Physica A* **238** 307.
18. Coolen A C C and Ruijgrok Th W 1988 *Phys. Rev. A* **38** 4253.
19. Nakano K 1972 *IEEE Trans. on Systems, Man, and Sybernetics* **SMC-2** 380.
20. Hopfield J J 1982 *PNAS* **79** 2554.
21. Coolen A C C and Sherrington D 1994 *Phys. Rev. Let.* **49** 1921.
22. Coolen A C C, Laughton S N and Sherrington D 1996 *Phys. Rev. B* **53** 8184.
23. Sherrington D and Kirkpatrick S 1975 *Phys. Rev. Lett.* **35** 1792.
24. Nishimori H 2001 *Statistical Physics of Spin Glasses and Information Processing: An Introduction* (Oxford: Oxford University Press).
25. Tanaka K and Horiguchi T 1997 *IEICE* **J80-A-12** 2217 (in Japanese).
26. Nishimori H and Wong K Y M 1999 *Phys. Rev. E* **60** 132.
27. Tanaka K 2002 *J. Phys. A: Math. Gen.* **35** R81.
28. Inoue J 2001 *Phys. Rev. E* **63** 046114.
29. Inoue J 2005 *Quantum Spin Glasses, Quantum Annealing, and Probabilistic Information Processing*, in *Quantum Annealing and Related Optimization Methods Lecture Notes in Physics* **679**, ed Das A and Chakrabarti B K (Berlin Heidelberg: Springer) p. 259.
30. Chakrabarti K, Dutta A and Sen P 1996 *Quantum Ising Phases and Transitions in Transverse Ising Models* (Heidelberg: Springer).
31. Sachdev S 1999 *Quantum Phase Transitions* (Cambridge: Cambridge University Press).
32. Sourlas N 1989 *Nature* **339** 693.
33. Inoue J, Saika Y and Okada M 2009 *J. Phys: Conference Series* **143** 012019.
34. Suzuki M 1966 *J. Phys. Soc. Japan* **21** 2140.
35. Elliot R J, Pfeuty P and Wood C 1970 *Phys. Rev. lett.* **25** 443.
36. Feynman R P and Hibbs A R 1965 *Quantum Mechanics and Path Integrals* (New York: McGraw-Hill).

37. Kleinert H 2009 *Path Integrals in Quantum Mechanics, Statistics, Polymer Physics, and Financial Markets* (Singapore: World Scientific).
38. Ozeki T and Okada M 2003 *J. Phys. A* **36** 11011.
39. Inoue J and Tanaka K 2002 *Phys. Rev. E* **65** 016125.
40. Ma Y Q and Gong C D 1992 *Phys. Rev. B* **45** 793.
41. Nishimori H and Nonomura Y 1996 *J. Phys. Soc. Japan* **65** 3780.
42. Chandra A K, Inoue J and Chakrabarti B K 2010 *Phys. Rev. E* **81** 021101.
43. Amit D, Gutfreund H and Somplolinsky H 1985 *Phys. Rev. Lett.* **55** 1530.
44. Inoue J 1996 *J. Phys. A* **29** 4815.
45. Shiino M and Fukai T *J. Phys. A* **25** L375.
46. Amari S and Maginu K 1988 *Neural Networks* **1** 63.
47. Okada M 1995 *Neural Networks* **8** 833.
48. Takahashi K and Matsuda Y 2010 *J. Phys. Soc. Japan* **79** 043712.
49. Das A, Sengupta K, Sen D and Chakrabarti B K 2006 *Phys. Rev. B* **74** 144423.
50. Das A 2010 *Phys. Rev. B* **82** 172402.
51. Das A and Chakrabarti B K 2008 *Rev. Mod. Phys.* **80** 1061.

A METHOD TO CONTROL ORDER OF PHASE TRANSITION: INVISIBLE STATES IN DISCRETE SPIN MODELS

RYO TAMURA

Institute for Solid State Physics, University of Tokyo,
5-1-5, Kashiwanoha, Kashiwa-shi, Chiba, Japan, 277-8581

SHU TANAKA

Department of Chemistry, University of Tokyo,
7-3-1, Hongo, Bunkyo-ku, Tokyo, Japan, 113-0033

NAOKI KAWASHIMA

Institute for Solid State Physics, University of Tokyo,
5-1-5, Kashiwanoha, Kashiwa-shi, Chiba, Japan, 277-8581

It is an important topic to investigate nature of the phase transition in wide area of science such as statistical physics, materials science, and computational science. Recently it has been reported the efficiency of quantum adiabatic evolution/quantum annealing for systems which exhibit a phase transition, and we cannot obtain a good solution in such systems. Thus, to control the nature of the phase transition has been also attracted attention in quantum information science. In this paper we review nature of the phase transition and how to control the order of the phase transition. We take the Ising model, the standard Potts model, the Blume-Capel model, the Wajnflasz-Pick model, and the Potts model with invisible states for instance. Until now there is no general method to avoid the difficulty of annealing method in systems which exhibit a phase transition. It is a challenging problem to propose a method how to erase or how to control the nature of the phase transition in the target system.

Keywords: Phase transition; Potts model with invisible states; Entropy effect; Quantum annealing.

1. Introduction

To study nature of the phase transition has been attracted attention in wide area of science such as materials science and statistical physics. If we change control parameters such as temperature, pressure, and external field, we can observe phase transition in many materials. In respective materials, a macroscopic ordered state (*e.g.* ferromagnetism, ferroelectoricity, superfluidity, and superconductivity) appears, which is characterized by each corresponding order parameter. The Ising model was proposed in 1925, which is regarded as the most fundamental magnetic model to study the nature of the phase transition from a microscopic viewpoint in statistical physics.[1] It is not too much to say that the Ising model triggers the development of statistical physics. Since the Ising model can be easily generalized because of simplicity, a number of generalized models have been proposed up to the present. The relation between the universality class and the symmetry which breaks at the transition point has been studied by analysis of these models. In general, the Hamiltonian of the Ising model with site-dependent external magnetic field h_i is expressed by

$$\mathcal{H}_{\text{Ising}} = -\sum_{i,j} J_{ij}\sigma_i^z\sigma_j^z - \sum_i h_i\sigma_i^z, \quad \sigma_i^z = +1, -1, \qquad (1)$$

where σ_i^z is a microscopic variable at the i-th site. Hereafter the g-factor and the Bohr magneton μ_B are set to be unity for simplicity. When $J_{ij} > 0$, the interaction between the i-th and j-th sites is ferromagnetic, whereas for $J_{ij} < 0$, the interaction is antiferromagnetic. We can adopt the Ising model for not only analysis of phase transition observed in real materials but also information science/technology. In information science/technology, a binary representation is a basic language. Then, an interdisciplinary science which is the interface between statistical physics and information science has been developed in terms of the Ising model. Actually some difficulties of information science/engineering were solved from a viewpoint of statistical physics.[2]

Optimization problem is a problem to find the minimum/maximum value of real-valued cost/gain function. Then, to solve an optimization problem corresponds to find the equilibrium state at finite temperature or the ground state of the Hamiltonian which expresses the target optimization problem. In many cases we can represent optimization problem by the Ising model or its generalized model. An algorithm which can solve optimization problem in a general way was proposed by Kirkpatrick. This method is called simulated annealing.[3,4] In general, energy landscape of optimization problems is complicated as random spin systems.[5–7] In the simulated annealing, we gradually decrease temperature and can obtain a not so bad or the best solution. A characteristic transition time at the temperature T is expressed $\tau \propto e^{\beta\Delta E}$, where β denotes the inverse temperature ($\beta = T^{-1}$) and ΔE represents a characteristic energy difference. For simplicity, the

Boltzmann constant k_B is set to be unity. Since the probability distribution of equilibrium state is almost flat at high temperature, τ becomes short. It was shown that by decreasing temperature slow enough, we can obtain the best solution of optimization problems by Geman and Geman.[8] Since the simulated annealing is easy to implement, this is adopted in many cases.[9]

The simulated annealing finds a not so bad solution or the best solution by making use of thermal fluctuation. In 1998, on the contrary, Kadowaki and Nishimori proposed an alternative method of simulated annealing, which is called quantum annealing.[10] In the quantum annealing, we decrease the quantum field (*i.e.* quantum fluctuation) instead of temperature in the simulated annealing.[11-15] By using the quantum annealing, we can succeed to obtain a better solution than the solution obtained by the simulated annealing in many cases.[16-22] Thus, the quantum annealing is expected as a powerful tool for optimization problems.[23-31] In order to improve this method more efficient, the quantum annealing from a viewpoint of statistical physics has been studied.[32] Annealing methods which are based on statistical physics seem to work well in all optimization problems since these methods are easily performed. However there is a weak point in both simulated annealing and quantum annealing. When we decrease temperature/quantum field across a phase transition point, we obtain not so good solution in general.[15] Especially, if a first-order phase transition occurs during annealing, we can not obtain the best solution definitely.[33] The situation is improved if a second-order phase transition takes place but some kind of critical slowing down exists.[34] Thus, it is an important issue to investigate how to erase the phase transition or how to change the order of the phase transition from first-order to second-order.

In this paper we focus on how to control the order of the phase transition. The organization of the rest of the paper is as follows. In Section 2, we explain nature of the phase transition taking as examples some fundamental models. In Section 3, we review how to change the order of the phase transition for preceding studies. We take the Blume-Capel model and the Wajnflasz-Pick model for instance. In Section 4, we consider the Potts model with invisible states. In Section 5, we summarize this paper and show future perspective.

2. Nature of the Phase Transition

Phase transition can be categorized into two types according to singularity of the free energy as a function of control parameter in the thermodynamic limit. If there is a singularity in differential coefficient of first order of free energy, a first-order phase transition occurs and then, energy and order parameter jump at the transition point. From this feature, a first-order phase transition is called discontinuous phase transition. On the other hand, if there is a singularity in differential coefficient of second order of free energy, a second-order phase transition takes place. In this case, energy and order parameter are continuous even at the transition point. Then, a second-order phase transition is called continuous phase transition as against discontinuous phase transition. When a second-order phase transition occurs, physical quantities near the transition point should be represented by power functions. The exponents of these functions are called critical exponents. A set of values of critical exponents corresponds to a universality class. Universality class has been investigated exhaustively by analytical methods and numerical methods such as Monte Carlo simulation. We can categorize a phase transition according to "*encyclopedia* of universality class".[a]

Here we explain the phase transition of the ferromagnetic Ising model with nearest-neighbor interaction under homogeneous external magnetic field. The Hamiltonian is given as

$$\mathcal{H}^{(2)}_{\text{Ising}} = -J \sum_{\langle i,j \rangle} \sigma_i^z \sigma_j^z - h \sum_i \sigma_i^z, \quad (J > 0), \quad \sigma_i^z = +1, -1. \quad (2)$$

Hereafter $\langle i, j \rangle$ denotes the nearest neighbor spin pair on the defined d-dimensional lattice, and we take J as the energy unit throughout this paper. The phase diagram of this model for $d \geq 2$ is depicted in Fig. 1 (a).[b] The horizontal and vertical axes in this figure are temperature T and external magnetic field h, respectively. The bold line and the circle in Fig. 1 (a) are the ferromagnetic phase and the critical point T_c. When we decrease temperature under zero field (see the line (i) in Fig. 1 (a)), a second-order phase transition with spontaneous twofold symmetry breaking occurs at the critical point T_c. In this model, the order parameter is magnetization defined as

$$m_{\text{Ising}} = \frac{1}{N} \sum_i \sigma_i^z, \quad (3)$$

where N is the number of spins. When the external magnetic field is zero, the behavior of magnetization is shown in Fig. 1 (b). On the other hand, when we change the external magnetic field with fixed temperature below

---

[a] In some cases, novel universality class is found. It should be noted that to explore new universality class itself is an important topic in statistical physics.
[b] On the one-dimensional lattice, there is no phase transition.

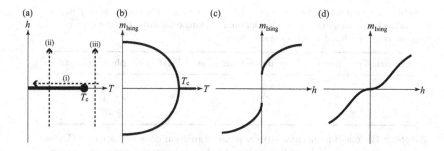

Fig. 1. (a) Phase diagram of the ferromagnetic Ising model for $d \geq 2$. The bold line represents ferromagnetic phase, and the circle denotes the critical point T_c. (b) Behavior of magnetization when we decrease temperature at $h = 0$ as the line (i) in (a). In actual, either the upper or lower branch is selected. This phenomenon is called spontaneous symmetry breaking. A second-order phase transition occurs at T_c. (c) Behavior of magnetization when we change external magnetic field at $T < T_c$ as the line (ii) in (a). A first-order phase transition occurs. (d) Behavior of magnetization when we change external magnetic field at $T > T_c$ as the line (iii) in (a). No phase transition happens.

T_c (see the line (ii) in Fig. 1 (a)), a first-order phase transition occurs and the magnetization jumps at $h = 0$ as depicted in Fig. 1 (c). When we change the external magnetic field with fixed temperature above T_c (see the line (iii) in Fig. 1 (a)), the magnetization behaves as a smooth function shown in Fig. 1 (d). When we sweep external magnetic field along the line (ii) in Fig. 1 (a) at finite speed, hysteresis curve is often observed because of existence of the metastable state.

Suppose we perform simulated/quantum annealing for systems which exhibit a first-order phase transition. When we decrease temperature/quantum field across the transition point, the state is trapped in the metastable states. Then we cannot obtain not so bad solution by annealing methods. Next we consider the case for second-order phase transition. Physical quantities converge to the equilibrium value slowly because of some kind of critical slowing down. This nature was studied recently, which is called the Kibble-Zurek mechanism.[34–38]

The relation between the order of the phase transition and the symmetry which breaks at the transition point has been considered by using the standard ferromagnetic Potts model.[39,40] The Hamiltonian of the ferromagnetic q-state Potts model is given as

$$\mathcal{H}_{\text{Potts}} = -J \sum_{\langle i,j \rangle} \delta_{s_i, s_j}, \quad (J > 0), \quad s_i = 1, \cdots, q. \quad (4)$$

Since the 2-state Potts model is equivalent to the Ising model, the q-state

Table 1. Relation between the order of the phase transition and q in the standard ferromagnetic Potts model on d-dimensional lattice. Note that q-fold symmetry breaks at the transition point.

Dimension d	Second-order phase transition	First-order phase transition
1	×[a]	×[a]
2	$q \leq 4$	$q > 4$
$d \geq 3$	$q = 1, 2$[b]	$q \geq 3$[b]

Note: [a] On one-dimensional lattice, phase transition does not occur. [b] There is no exact result of boundary value of q between second-order phase transition and first-order phase transition. It is true that q-state Potts model on d-dimensional ($d \geq 3$) lattice for $q = 1, 2$ exhibits a second-order phase transition whereas a first-order phase transition takes place in that model for $q \geq 3$.

Potts model is regarded as a straightforward extension of the Ising model. The Potts model is used for analysis of coloring problems and clustering problems,[19,20] and plays an important role for not only statistical physics but also information science. In this model on one-dimensional lattice, there is no phase transition at finite temperature for arbitrary q as well as the Ising model.

It is convenient to introduce another representation of Kronecker's delta as

$$\delta_{\mu,\nu} = \frac{1 + (q-1)\mathbf{e}^\mu \cdot \mathbf{e}^\nu}{q}, \tag{5}$$

where $\mathbf{e}^\mu$ ($\mu = 1, \cdots, q$) represents q unit vectors pointing in the q symmetric direction of a hyper-tetrahedron in $q - 1$ dimensions.[40] Then the order parameter of this model can be defined as

$$\mathbf{m}_{\text{Potts}} = \frac{1}{N} \sum_i \mathbf{e}^{s_i}. \tag{6}$$

This model on two-dimensional lattice exhibits a second-order phase transition with q-fold symmetry breaking for $q \leq 4$ whereas a first-order phase transition with q-fold symmetry breaking for $q > 4$. In a similar way, the relation between the order of the phase transition and the symmetry which breaks at the transition point for $d > 2$ is also investigated as shown in Table 1.[41]

3. Preceding Models

In this section we review two famous preceding models. Both of two models are some kind of generalized Ising model.

3.1. Blume-Capel model

We explain nature of the phase transition of the Blume-Capel model.[42,43] The Hamiltonian of this model is given as

$$\mathcal{H}_{\text{BC}} = -J \sum_{\langle i,j \rangle} t_i t_j - D \sum_i t_i^2, \quad (J > 0), \quad t_i = +1, 0, -1. \tag{7}$$

Here we refer to $t_i = 0$ as vacancy. At $D = -\infty$, vacancy is suppressed and the state of each spin is $t_i = +1$ or -1, whereas at $D = +\infty$, all spins become vacancies. This model can represent magnetic lattice gas or annealed diluted Ising model, where D corresponds to the chemical potential. This model has been widely used analysis of multicritical phenomena in metallic alloys and liquid crystals etc.[44-47]

Schematic phase diagram of the Blume-Capel model is depicted in Fig. 2. The solid and dotted curves are second-order phase transition and first-order phase transition, respectively. The circle between the solid and dotted curves in Fig. 2 represents the tricritical point (T_c, D_c). Thus the Blume-Capel model is a fundamental model which has a tricritical point. In this model we can change the order of the phase transition by controlling the chemical potential D. When we decrease temperature for fixed $D(< D_c)$, a second-order phase transition occurs, whereas a first-order phase transition takes place when we decrease temperature for fixed $D(> D_c)$. It should be noted that the ground state is also changed when we control D. Then this type of control method for changing the order of the phase transition is not suitable for annealing methods.

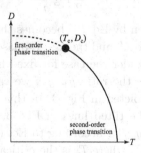

Fig. 2. Schematic phase diagram of the Blume-Capel model. The solid and dotted curves represent second-order phase transition and first-order phase transition, respectively. The circle indicates the tricritical point (T_c, D_c).

3.2. Wajnflasz-Pick model

Next we review nature of the phase transition of the Wajnflasz-Pick model.[48] The Hamiltonian of the Wajnflasz-Pick model is given as

$$\mathcal{H}_{\text{WP}} = -J \sum_{\langle i,j \rangle} s_i s_j - h \sum_i s_i, \quad s_i = \underbrace{+1, \cdots, +1}_{g_+}, \underbrace{-1, \cdots, -1}_{g_-}, \quad (8)$$

where g_+ and g_- are the number of $+1$-state and that of -1-state, respectively, and we assume $J > 0$. Note that this model for $g_+ = g_- = 1$ corresponds to the standard Ising model. We can transform this Hamiltonian into the following Hamiltonian at finite temperature T:

$$\mathcal{H}'_{\text{WP}} = -J \sum_{\langle i,j \rangle} \sigma_i^z \sigma_j^z - (h - \frac{T}{2} \log \frac{g_+}{g_-}) \sum_i \sigma_i^z, \quad \sigma_i^z = +1, -1. \quad (9)$$

Note that the number of $+1$-state and that of -1-state are unity. These two Hamiltonians are equivalent since the partition functions of these Hamiltonians are the same, i.e. $\text{Tr}\, e^{-\beta \mathcal{H}_{\text{WP}}} = \text{Tr}\, e^{-\beta \mathcal{H}'_{\text{WP}}}$. The second term of Eq. (9) consists of the original external magnetic field h and temperature-dependent part. The latter comes from the entropy effect of the bias of g_+ and g_-. When $g_+ = g_-$, this term disappears, and the Hamiltonian given by Eq. (9) becomes the standard Ising model. From this fact, the temperature-dependent external magnetic field can be regarded as the entropy-induced internal field. The Wajnflasz-Pick model has been adopted for analysis of phase transition in spin-crossover materials.[49-53]

We can analyze phase transition in this model by using the phase diagram of the standard Ising model. Here, the coefficient of second term in Eq. (9) is represented as

$$h' := h - \frac{T}{2} \log \frac{g_+}{g_-}. \quad (10)$$

Then the Hamiltonian given by Eq. (9) becomes the standard Ising model with external magnetic field h', and the $h' - T$ phase diagram of this model is shown in Fig. 3. We consider the case for fixed finite external magnetic field $h = h_0$. If we change the ratio g_+/g_-, we can control the order of the phase transition as depicted in Fig. 3. In this model, to change temperature corresponds to the tilted lines in Fig. 3, and a slope of trajectory changes by the ratio g_+/g_- according to Eq. (10). If we set the ratio $g_+/g_- = \exp(2h_0/T_c) =: g^*$, where T_c is the critical point of the standard Ising model, a second-order phase transition occurs when we decrease temperature. If the ratio g_+/g_- is smaller than g^*, no phase transition occurs, whereas a first-order phase transition takes place when g_+/g_- is larger

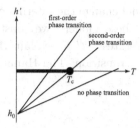

Fig. 3. The bold line and the circle represent ferromagnetic phase and the critical point T_c of the Ising model. The thin lines correspond to changing the temperature at fixed external magnetic field h_0. The gradients of these lines are equal to $\log(g_+/g_-)/2$.

than g^*. Thus, the Wajnflasz-Pick model is a standard model where we can change the order of the phase transition without changing the ground state by just controlling the ratio g_+/g_-. It should be noted that the all phase transitions in this model do not accompany the twofold symmetry breaking which is the characteristic property of the standard Ising model, except for $h_0 = 0$ and $g_+ = g_-$ (i.e. the standard ferromagnetic Ising model without external field).

4. Potts Model with Invisible States

In Section 2, we reviewed nature of the phase transition in the standard ferromagnetic Potts model.[39,40] It has been considered that the relation between the order of the phase transition and the symmetry which breaks at the transition point was investigated completely. The standard ferromagnetic Potts model has actually succeeded to analyze a phase transition with discrete symmetry breaking appeared in real materials and complicated theoretical models. Very recently, however, some phase transitions which are not consistent with the nature of the phase transition in the standard Potts model were reported, although these phase transitions accompany discrete symmetry breaking.[54-57] For instance, a first-order phase transition with threefold symmetry breaking occurs on two-dimensional lattice. According to Table 1, the 3-state ferromagnetic Potts model on two-dimensional lattice should exhibit a second-order phase transition with threefold symmetry breaking. This behavior is controversial feature. In order to understand what happens in such phase transitions, we should propose a new model.

As we mentioned in Section 3, we can control the order of the phase transition by changing the chemical potential of vacancy or changing the bias of the number of states. Roughly speaking, the Blume-Capel model and the Wajnflasz-Pick model change the internal energy and the entropy, respectively. Motivated by these models, we constructed a new model – Potts model with invisible states[58-60] to explain the nature of the above mentioned intriguing phase transition. The Hamiltonian of this model is given as

$$\mathcal{H}_{\text{PI}} = -J \sum_{\langle i,j \rangle} \delta_{t_i,t_j} \sum_{\alpha=1}^{q} \delta_{t_i,\alpha}, \quad t_i = 1, \cdots, q, q+1, \cdots, q+r, \quad (11)$$

where the states $1 \leq t_i \leq q$ and $q+1 \leq t_i \leq q+r$ are referred to as colored states and invisible states, respectively, and we assume $J > 0$. Hereafter we refer to this model as (q,r)-state Potts model. Obviously, this model for $r = 0$ corresponds to the standard Potts model, and then, this model is regarded as the straight forward extension of the standard Potts model.

First, we consider two spin case for simplicity i.e. $\mathcal{H}_{\text{PI}} = -J\delta_{t_1,t_2} \sum_{\alpha=1}^{q} \delta_{t_1,\alpha}$. When $1 \leq t_1 = t_2 \leq q$, the energy becomes $-J$, otherwise the energy becomes zero. Notice that even when $t_1 = t_2$, the energy becomes zero if $q+1 \leq t_1 = t_2 \leq q+r$, which differs from the case of the standard ferromagnetic Potts model. The energy structure of this model is depicted in Fig. 4. The number of ground states is q and the number of excited states is $(q+r)^2 - q$, whereas the number of excited states is $q^2 - q$ for $r = 0$ (standard Potts model). Only the number of excited states increases by adding the invisible states into the standard Potts model. It should be noted that the number of ground states of the model given by Eq. (11) and that of the standard Potts model are the same.

At finite temperature T, we can transform this Hamiltonian into the

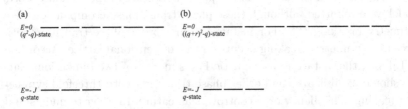

Fig. 4. (a) Energy structure of q-state Potts model for $N = 2$. The number of ground states and that of excited states are q and $q^2 - q$, respectively. (b) Energy structure of (q,r)-state Potts model for $N = 2$. The number of ground states and that of excited states are q and $(q+r)^2 - q$, respectively.

following form as well as the Wajnflasz-Pick model:

$$\mathcal{H}'_{\text{PI}} = -J\sum_{\langle i,j\rangle}\delta_{s_i,s_j}\sum_{\alpha=1}^{q}\delta_{s_i,\alpha} - T\log r\sum_{i}\delta_{s_i,0}, \quad s_i = 0,1,\cdots,q. \quad (12)$$

Here we rename the label of invisible states from $q + 1 \leq t_i \leq q + r$ to $s_i = 0$. The partition functions of both Hamiltonians are the same: $\text{Tr}\,e^{-\beta\mathcal{H}_{\text{PI}}} = \text{Tr}\,e^{-\beta\mathcal{H}'_{\text{PI}}}$. The second term of this Hamiltonian is regarded as the chemical potential of invisible states and comes from the entropy effect of the number of invisible states r.

4.1. Mean-field analysis

We study phase transition in the (q,r)-state Potts model by the Bragg-Williams approximation as well as the standard ferromagnetic q-state Potts model.[41] Since the number of ground states of this model is q, if a phase transition takes place, q-fold symmetry breaks at the transition point T_c. Let w_α be the fraction of the α-th state ($0 \leq \alpha \leq q$). Here we use the notation of the Hamiltonian given by Eq. (12) and then the label of the invisible state is zero. We assume that the first state is selected below the transition point T_c after q-fold symmetry breaks. Then $\{w_\alpha\}$ can be represented as

$$\begin{cases} w_0 = y \\ w_1 = \frac{1}{q}(1-y)[1+(q-1)x] \\ w_\alpha = \frac{1}{q}(1-y)(1-x) \quad (2\leq\alpha\leq q) \end{cases} \quad (13)$$

where $0 \leq x, y \leq 1$. The fraction of the invisible states is represented by y, and x indicates the degree of ordering. When $x = 0$, the system is paramagnetic state, whereas when $x = 1$, the system is completely ordered state. It is a natural condition that w_α's for $2 \leq \alpha \leq q$ are the same. The internal energy and the entropy in the level of the Bragg-Williams approximation are given by

$$E^{\text{BW}}(x,y) = -w_0 T\log r - \frac{zJ}{2}\sum_{\alpha=1}^{q}w_\alpha^2, \quad (14)$$

$$S^{\text{BW}}(x,y) = -\sum_{\alpha=0}^{q}w_\alpha\log w_\alpha, \quad (15)$$

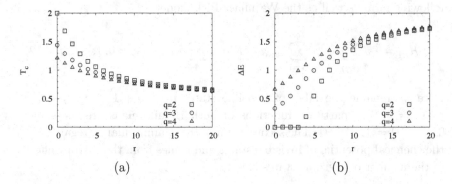

Fig. 5. (a) Transition temperature as a function of the number of invisible states r for $z = 4$. (b) Latent heat as a function of r for $z = 4$. Both results are obtained by the Bragg-Williams approximation.

respectively. The parameter z in Eq. (14) is the number of the nearest-neighbor sites. As a result, the free energy can be written as

$$F^{\text{BW}}(x,y) = E^{\text{BW}}(x,y) - TS^{\text{BW}}(x,y)$$
$$= -\frac{zJ(1-y)^2}{2q}[(q-1)x^2 + 1] + yT\log\frac{y}{r}$$
$$+ (1-y)T\left[\frac{1+(q-1)x}{q}\log\frac{1+(q-1)x}{1-x} + \log\frac{(1-y)(1-x)}{q}\right]. \quad (16)$$

From this free energy, the transition temperature and the latent heat can be obtained by numerical calculation. The number of invisible states r dependencies of the transition temperature and the latent heat for $q = 2, 3$, and 4 are shown in Fig. 5.

The latent heat expresses the energy difference between metastable state and stable state at the transition temperature. By the definition, if the latent heat is a finite value, the internal energy jumps at the transition temperature. Thus a finite latent heat indicates that a first-order phase transition occurs. From Fig. 5 (b), the $(2,r)$-state Potts model for $r \leq 3$ exhibits a second-order phase transition and others have a first-order phase transition. As the number of invisible states r increases, the transition temperature decreases and the latent heat increases. This result is quite natural since the invisible states contribute to the entropy. To clarify the effect of the invisible states, we compare with the transition temperature and the latent heat for $r = 0$ (standard Potts model). The transition temperature

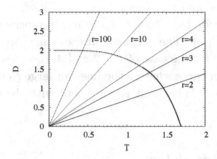

Fig. 6. Phase diagram of the Blume-Emery-Griffiths model for $J = J'$ obtained by the Bragg-Williams approximation for $z = 4$. The thick solid and dotted curves indicate second-order phase transition and first-order phase transition. The thin lines correspond to decreasing temperature for $r = 2, 3, 4, 10,$ and 100 from bottom to top. The gradients of these lines are $\log r$.

and the latent heat for $r = 0$ are obtained as

$$\begin{cases} T_c^{\text{BW}}(q = 2, r = 0) = \frac{zJ}{2} \\ \Delta E^{\text{BW}}(q = 2, r = 0) = 0 \end{cases}, \tag{17}$$

$$\begin{cases} T_c^{\text{BW}}(q \geq 3, r = 0) = \frac{zJ}{2\log(q-1)} \left(\frac{q-2}{q-1}\right) \\ \Delta E^{\text{BW}}(q \geq 3, r = 0) = \frac{zJ(q-2)^2}{2q(q-1)} \end{cases}. \tag{18}$$

The transition temperature and the latent heat for $r = 0$ and $z = 4$ are also shown in Fig. 5.

We also consider the relation between the $(2,r)$-state Potts model and the Blume-Emery-Griffiths model[61] which is an extended model of the Blume-Capel model. The Hamiltonian of the Blume-Emery-Griffiths model is given by

$$\mathcal{H}_{\text{BEG}} = -\frac{J}{2}\sum_{\langle i,j \rangle} t_i t_j - \frac{J'}{2}\sum_{\langle i,j \rangle} t_i^2 t_j^2 - D\sum_i (1 - t_i^2), \quad t_i = +1, 0, -1, \tag{19}$$

where the biquadratic interaction (the second term) is introduced into the Blume-Capel model. The Blume-Emery-Griffiths model for $J = J'$ and $D = T \log r$ is equivalent to the $(2,r)$-state Potts model. In order to investigate nature of the phase transition of this model, we obtain the phase diagram of the model for $J = J'$ and unfixed D by the Bragg-Williams approximation for $z = 4$.

The phase diagram of the Blume-Emery-Griffiths model for $J = J'$ is depicted in Fig. 6. The thick solid and dotted curves indicate second-order phase transition and first-order phase transition, respectively. Here we consider the case for $D = T \log r$. To change temperature corresponds to the thin lines in Fig. 6. The gradients of these lines are $\log r$. When we decrease temperature for the case of $r = 2$ and 3, the lines cross a second-order phase transition curve whereas the lines for large r cross a first-order phase transition curve. This result is consistent with the result shown in Fig. 5.

4.2. Monte Carlo simulation

In the previous subsection we studied the order of the phase transition and the latent heat in the (q,r)-state Potts model by the Bragg-Williams approximation. These results correspond to the infinite-dimensional version of the (q,r)-state Potts model. Then in order to consider nature of the phase transition in the (q,r)-state Potts model on finite-dimensional lattice, we study thermodynamic properties of this model on the $L \times L (= N)$ square lattice by Monte Carlo simulation.

First we calculate temperature dependencies of the order parameter m_{Potts}, the density of invisible states ρ_{inv}, the internal energy E, and the specific heat C for $(q,r)=(4,20)$. The definition of ρ_{inv} is defined by

$$\rho_{\text{inv}} = \frac{1}{N} \sum_i \sum_{\alpha=q+1}^{q+r} \delta_{t_i,\alpha}. \tag{20}$$

It should be noted that we use the order parameter of the standard ferromagnetic q-state Potts model given by Eq. (6). These quantities as functions of temperature are shown in Fig. 7. Figure 7 (a) shows temperature dependency of the order parameter $|m_{\text{Potts}}|^2$, which indicates fourfold symmetry breaks at the transition temperature. The fraction of the invisible states as a function of temperature is shown in Fig. 7 (b). As the temperature increases, the fraction of invisible states increases. At $T = \infty$, this value becomes $r/(q+r) = 0.833\cdots$ by the definition whereas this value becomes zero at $T = 0$. Figure 7 (c) and (d) display the internal energy and the specific heat, respectively. The specific heat is not divergent behavior but has a finite large value at the transition temperature. This result indicates that a first-order phase transition takes place.

In order to confirm that this phase transition is of first-order, we calculate energy histogram at the temperature $T_c(L)$ where the specific heat has the maximum value. The temperature $T_c(L)$ is obtained by reweighting

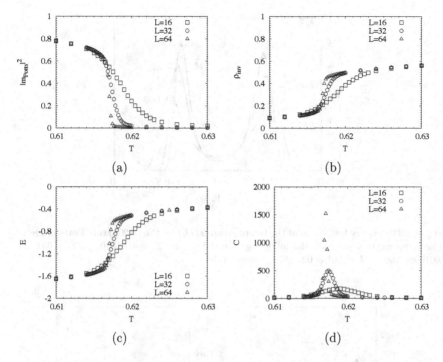

Fig. 7. Temperature dependencies of (a) the order parameter, (b) the fraction of the invisible states, (c) the internal energy, and (d) the specific heat for the (4,20)-state Potts model. The error bars are omitted for clarity since they are smaller than the symbol size.

method.[62,63] The energy histogram $P(E)$ is defined as

$$P(E) = W(E)e^{-\beta E}, \tag{21}$$

where $W(E)$ is the number of states in energy E. Figure 8 shows the energy histogram at $T_c(L)$ for $L = 32$, 64, and 96. A bimodal distribution is observed, which is a characteristic behavior of system which exhibits a first-order phase transition. As the number of spins increases, the two peaks become sharp. In the thermodynamic limit, these two peaks are expected to be the delta functions.

Next we take finite-size scaling in order to determine the transition temperature and the latent heat in the thermodynamic limit. We adopt the following functions:[64]

$$T_c(L) = aL^{-d} + T_c, \tag{22}$$

$$C_{\max}(L) \propto \frac{(\Delta E)^2 L^d}{4T_c^2}, \tag{23}$$

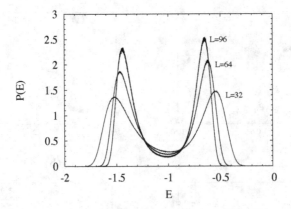

Fig. 8. The energy histogram at the temperature $T_c(L)$ for the (4,20)-state Potts model. The temperatures are used the following values $T_c(L = 32) = 0.61740$, $T_c(L = 64) = 0.61698$, and $T_c(L = 96) = 0.616894$, respectively.

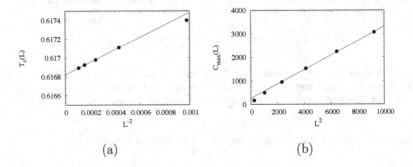

Fig. 9. (a) The temperature $T_c(L)$ as a function of L^{-2} for the (4,20)-state Potts model. The dotted line is a fitting line given by Eq. (22). The intercept of the fitting line indicates the transition point T_c. (b) The maximum value of the specific heat as a function of L^2 for the (4,20)-state Potts model. The dotted line is a fitting line given by Eq. (23).

where $C_{\max}(L)$ and ΔE are the maximum value of the specific heat for $L \times L$ system and the latent heat in the thermodynamic limit, respectively. In this case, we adopt $d = 2$. Figure 9 (a) shows $T_c(L)$ as a function of L^{-2}. The intercept of the dotted line corresponds to the transition temperature T_c. The maximum value of the specific heat $C_{\max}(L)$ as a function of L^2 is depicted in Fig. 9 (b). From these results, the transition temperature and the latent heat are obtained as $T_c = 0.61683(1)$ and $\Delta E = 0.68(2)$, respectively. This fact indicates that we succeeded to construct model which

Table 2. The transition temperature and the latent heat for several (q,r) on the square lattice.

(q,r)	Latent heat	Transition temperature
(2,0)[a]	0	1.13459
(2,30)[b]	1.02(2)	0.57837(1)
(2,32)[c]	1.23(2)	0.56857(1)
(3,0)[a]	0	0.994973
(3,25)[b]	0.81(2)	0.59630(1)
(3,27)[c]	1.05(2)	0.58513(1)
(4,0)[a]	0	0.910239
(4,20)[b]	0.68(2)	0.61683(1)
(4,22)[c]	0.87(2)	0.60396(1)

Note: [a] These results were obtained exactly. [b] These results were first obtained in Ref. 58. [c] These results were obtained in Ref. 59.

exhibits a first-order phase transition with fourfold symmetry breaking on the two-dimensional lattice. By the same procedure, we can obtain the transition temperature and the latent heat for several (q,r) (see Table 2).

As the number of invisible states increases, the transition temperature decreases and the latent heat increases, which is qualitatively consistent with the result obtained by the Bragg-Williams approximation. Thus, we conclude that invisible state in the Potts model stimulates a first-order phase transition. From this, unfortunately, this model is not useful for the optimization problems. Recently, nature of the phase transition in the (q,r)-state Potts model has been confirmed by a number of researchers.[65–67]

4.3. *Another representation of the Potts model with invisible states*

So far, we showed thermodynamic properties and phase transition of the (q,r)-state Potts model. This model is regarded as a straightforward extension of the standard q-state Potts model. In order to make it more clear, we show another representation of the Potts model with invisible states. First we consider the standard q-state Potts model. Let $\vec{S}_i$ be a state vector at the i-th site. The state vector is q-dimensional binary vector. Only one element in this vector is one and the others are zero. The position of one corresponds to the state. For example, when $\vec{S}_i = {}^\mathrm{T}(0,0,1,0,\cdots,0)$, the state of the i-th site is the third state. Here ${}^\mathrm{T}\vec{v}$ represents the transpose of the vector $\vec{v}$. We can express the Hamiltonian by matrix representation

as follows:
$$\mathcal{H}_{\text{Potts}} = -J\sum_{\langle i,j\rangle} \delta_{s_i,s_j} = -\sum_{\langle i,j\rangle} {}^{\text{T}}\vec{S}_i \hat{J}_{\text{Potts}} \vec{S}_j, \qquad (24)$$

where $\hat{J}_{\text{Potts}}$ is a q-by-q diagonal matrix whose elements are expressed as
$$\hat{J}_{\text{Potts}} = \text{diag}(\underbrace{J,\cdots,J}_{q}). \qquad (25)$$

Next we consider such a representation of the (q,r)-state Potts model. Let $\vec{T}_i$ be a $(q+r)$-dimensional indicator vector. The Hamiltonian of the Potts model with invisible states is represented as
$$\mathcal{H}_{\text{PI}} = -J\sum_{\langle i,j\rangle} \delta_{t_i,t_j} \sum_{\alpha=1}^{q} \delta_{t_i,\alpha} = -\sum_{\langle i,j\rangle} {}^{\text{T}}\vec{T}_i \hat{J}_{\text{PI}} \vec{T}_j, \qquad (26)$$

where $\hat{J}_{\text{PI}}$ is a $(q+r)$-by-$(q+r)$ diagonal matrix whose elements are expressed as
$$\hat{J}_{\text{PI}} = \text{diag}(\underbrace{J,\cdots,J}_{q},\underbrace{0,\cdots,0}_{r}). \qquad (27)$$

We can construct more generalized model by using the matrix representation by introducing the off-diagonal elements like the clock model, for instance. It is an important topic to control the order of the phase transition by changing the form of the interaction matrix $\hat{J}$.

5. Conclusion and Future Perspective

In this paper we reviewed nature of first-order and second-order phase transitions in general by taking the Ising model for instance. We demonstrated how to change the order of the phase transition by using the Blume-Capel model and the Wajnflasz-Pick model. In the Blume-Capel model, we can change the order of the phase transition by changing the chemical potential of vacancy. In this model, the ground state changes when we control the chemical potential. In the Wajnflasz-Pick model, on the other hand, by changing the bias of the number of states, the order of the phase transition can be changed without changing the ground state. By introducing the effects in both models into the standard Potts model, we constructed a generalized Potts model – Potts model with invisible states. The invisible state corresponds to the vacancy as well as the Blume-Capel model and the number of the invisible states contributes to the entropy as well as the Wajnflasz-Pick model. We can control the order of the phase transition by just changing the number of invisible states without variation of ground state. Then as the number of invisible states increases, the transition temperature decreases and the latent heat increases.

To change the order of the phase transition is an important topic for not only statistical physics but also computational science and quantum information theory (especially, quantum annealing or quantum adiabatic computation). Simulated annealing and quantum annealing which are based on (quantum) statistical physics do not work well in systems which exhibit a first-order phase transition. If a second-order phase transition takes place, a situation becomes not so bad, but a kind of critical slowing down problem remains. Phase transition in systems is obstacle in optimization problems whenever we use some kind of annealing procedures. Then it is anticipated to propose a method to control the order of the phase transition or to erase the phase transition if possible. Unfortunately, the invisible state in the Potts model stimulates a first-order phase transition and is not useful for annealing methods in the present stage. However our proposed method to generalize models is quite general and simple. Thus it is expected that how to erase the phase transition or how to change the order of the phase transition will be found in a similar way.

Acknowledgements

The authors are grateful to Jie Lou, Yoshiki Matsuda, Seiji Miyashita, Takashi Mori, Yohsuke Murase, Taro Nakada, Masayuki Ohzeki, Per Arne Rikvold, and Eric Vincent for their valuable comments. R.T. is partly supported by Global COE Program "the Physical Sciences Frontier", MEXT, Japan. S.T. is partly supported by Grand-in-Aid for JSPS Fellows (23-7601). The present work is financially supported by MEXT Grant-in-Aid for Scientific Research (B) (22340111), and for Scientific Research on Priority Areas "Novel States of Matter Induced by Frustration" (19052004), and by Next Generation Supercomputing Project, Nanoscience Program, MEXT, Japan. The computation in the present work was performed on computers at the Supercomputer Center, Institute for Solid State Physics, University of Tokyo.

References

1. E. Ising, *Z. Physik* **31**, 253 (1925).
2. H. Nishimori, *Statistical physics of spin glasses and information processing: an introduction* (Oxford University Press, 2001).
3. S. Kirkpatrick, C. D. Gelatt Jr., and M. P. Vecchi, *Science* **220**, 671 (1983).

4. S. Kirkpatrick, *J. Stat. Phys.* **34**, 975 (1984).
5. M. Mézard, G. Parisi, and M. A. Virasoro, *Spin Glass Theory and Beyond* (World Scientific, 1987).
6. K. H. Fischer and J. A. Hertz, *Spin Glasses* (Cambridge University Press, 1993).
7. A. P. Young, *Spin Glasses and Random Fields* (World Scientific, 1998).
8. S. Geman and D. Geman, *IEEE Transactions on Pattern Analysis and Machine Intelligence* **6**, 721 (1984).
9. P. J. M. Van Laarhoven and E. H. L. Aarts, *Simulated Annealing: Theory and Applications* (D Reidel, 1987).
10. T. Kadowaki and H. Nishimori, *Phys. Rev. E* **58**, 5355 (1998).
11. J. Brooke, D. Bitko, T. F. Rosenbaum, and G. Aeppli, *Science* **284**, 779 (1999).
12. E. Farhi, J. Goldstone, S. Gutmann, J. Lapan, A. Lundgren, and D. Preda, *Science* **292**, 472 (2001).
13. G. E. Santoro, R. Martoňák, E. Tosatti, and R. Car, *Science* **295**, 2427 (2002).
14. S. Suzuki and M. Okada, *J. Phys. Soc. Jpn.* **74**, 1649 (2005).
15. A. Das and B. K. Chakrabarti, *Quantum Annealing and Related Optimization Methods* (Springer, 2005).
16. R. Martoňák, G.E. Santoro, and E. Tosatti, *Phys. Rev. E* **70**, 057701 (2004).
17. D. A. Battaglia, G. E. Santoro, and E. Tosatti, *Phys. Rev. E* **71**, 066707 (2005).
18. S. Tanaka and S. Miyashita, *J. Magn. Magn. Mater.* **310**, e468 (2007).
19. K. Kurihara, S. Tanaka, and S. Miyashita, *Proceedings of The 25th Conference on Uncertainty in Artificial Intelligence* (2009).
20. I. Sato, K. Kurihara, S. Tanaka, H. Nakagawa, and S. Miyashita, *Proceedings of The 25th Conference on Uncertainty in Artificial Intelligence* (2009).
21. S. Morita, S. Suzuki, and T. Nakamura, *Phys. Rev. E* **79**, 065701(R) (2009).
22. J. Inoue, Y. Saika, and M. Okada, *Lecture Note in Physics "Quantum quenching, annealing, and computation"* (Springer) **802**, 283 (2010).
23. A. Das and B. K. Chakrabarti, *Rev. Mod. Phys.* **80**, 1061 (2008).
24. S. Miyashita, S. Tanaka, H. de Raedt, and B. Barbara, *J. Phys.: Conf. Ser.* **143**, 012005 (2009).
25. S. Tanaka and S. Miyashita, *Phys. Rev. E* **81**, 051138 (2010).
26. S. Tanaka, M. Hirano, and S. Miyashita, *Lecture Note in Physics "Quantum quenching, annealing, and computation"* (Springer) **802**, 215 (2010).
27. A. K. Chandra, A. Das, J. Inoue, and B. K. Chakrabarti, *Lecture Note in Physics "Quantum quenching, annealing, and computation"* (Springer) **802**, 235 (2010).
28. M. Ohzeki and H. Nishimori, *J. Comp. and Theor. Nanoscience* **8**, 963 (2011).
29. S. Tanaka, M. Hirano, and S. Miyashita, *Physica E* **43**, 766 (2010).
30. S. Tanaka, R. Tamura, I. Sato, and K. Kurihara, to appear in *Kinki University Quantum Computing Series: "Summer School on Diversities in Quantum Computation/Information"*.

31. S. Tanaka, to appear in *proceedings of Kinki University Quantum Computing Series: "Symposium on Quantum Information and Quantum Computing"* (2011).
32. M. Ohzeki, *Phys. Rev. Lett.* **105**, 050401 (2010).
33. A. P. Young, S. Knysh, and V. N. Smelyanskiy, *Phys. Rev. Lett.* **104**, 020502 (2010).
34. S. Suzuki, *Lecture Note in Physics "Quantum quenching, annealing, and computation"* (Springer) **802**, 114 (2010).
35. T. W. B. Kibble, *J. Phys. A* **9**, 1387 (1976).
36. T. W. B. Kibble, *Phys. Rep.* **67**, 183 (1980).
37. W. H. Zurek, *Nature* **317**, 505 (1985).
38. W. H. Zurek, *Acta. Phys. Pol. B* **24**, 1301 (1993).
39. R. B. Potts, *Proc. Camb. Phil. Soc.* **48**, 106 (1952).
40. F. Y. Wu, *Rev. Mod. Phys.* **54**, 235 (1982).
41. T. Kihara, Y. Midzuno, and T. Shizume, *J. Phys. Soc. Jpn.* **9**, 681 (1954).
42. M. Blume, *Phys. Rev.* **141**, 517 (1966).
43. H. W. Capel, *Physica* **32**, 966 (1966).
44. P. E. Cladis, R. K. Bogardus, W. B. Daniels, and G. N. Taylor, *Phys. Rev. Lett.* **39**, 720 (1977).
45. A. Blatter and M. von Allmen, *Phys. Rev. Lett.* **54**, 2103 (1985).
46. J. Zhang, X. Peng, A. Jonas, and J. Jonas, *Biochemistry* **34**, 8631 (1995).
47. S. Rastogi, G. W. H. Höhne, and A. Keller, *Macromolecules* **32**, 8897 (1999).
48. J. Wajnflasz and R. Pick, *J. Phys. Colloq. France* **32**, C1 (1971).
49. R. Zimmermann, *J. Phys. Chem. Sol.* **44**, 151 (1983).
50. K. Boukheddaden, I. Shteto, B. Hoo, and F. Varret, *Phys. Rev. B* **62**, 14806 (2000).
51. S. Miyashita and N. Kojima, *Prog. Theor. Phys.* **109**, 729 (2003).
52. M. Nishino, S. Miyashita, and K. Boukheddaden, *J. Chem. Phys.* **118**, 4594 (2003).
53. H. Tokoro, S. Miyashita, K. Hashimoto, and S. Ohkoshi, *Phys. Rev. B* **73**, 172415 (2006).
54. R. Tamura and N. Kawashima, *J. Phys. Soc. Jpn.* **77**, 103002 (2008).
55. E. M. Stoudenmire, S. Trebst, and L. Balents, *Phys. Rev. B* **79**, 214436 (2009).
56. S. Okumura, H. Kawamura, T. Okubo, and Y. Motome, *J. Phys. Soc. Jpn.* **79**, 114705 (2010).
57. R. Tamura and N. Kawashima, *J. Phys. Soc. Jpn.* **80**, 074008 (2011).
58. R. Tamura, S. Tanaka, and N. Kawashima, *Prog. Theor. Phys.* **124**, 381 (2010).
59. S. Tanaka, R. Tamura, and N. Kawashima, *J. Phys.: Conf. Ser.* **297**, 012022 (2011).
60. S. Tanaka and R. Tamura, *J. Phys.: Conf. Ser.* **320**, 012025 (2011).
61. M. Blume, V. J. Emery, and R. B. Griffiths, *Phys. Rev. A* **4**, 1071 (1971).
62. A. M. Ferrenberg and R. H. Swendsen, *Phys. Rev. Lett.* **61**, 2635 (1988).
63. A. M. Ferrenberg and R. H. Swendsen, *Phys. Rev. Lett.* **63**, 1195 (1989).
64. M. S. S. Challa, D. P. Landau, and K. Binder, *Phys. Rev. B* **34**, 1841 (1986).

65. A. C. D. van Enter, G. Iacobelli, and S. Taati, *arXiv*:1106.5907 (2011).
66. A. C. D. van Enter, G. Iacobelli, and S. Taati, *arXiv*:1109.0189 (2011).
67. T. Mori, *arXiv*:1111.1046 (2011).

239

QUANTUM ANNEALING AND QUANTUM FLUCTUATION EFFECT IN FRUSTRATED ISING SYSTEMS

SHU TANAKA

Department of Chemistry, University of Tokyo,
7-3-1, Hongo, Bunkyo-ku, Tokyo, 113-0033, Japan
E-mail: shu-t@chem.s.u-tokyo.ac.jp

RYO TAMURA

Institute for Solid State Physics, University of Tokyo,
5-1-5, Kashiwanoha, Kashiwa-shi, Chiba, 277-8501, Japan
E-mail: r.tamura@issp.u-tokyo.ac.jp

Quantum annealing method has been widely attracted attention in statistical physics and information science since it is expected to be a powerful method to obtain the best solution of optimization problem as well as simulated annealing. The quantum annealing method was incubated in quantum statistical physics. This is an alternative method of the simulated annealing which is well-adopted for many optimization problems. In the simulated annealing, we obtain a solution of optimization problem by decreasing temperature (thermal fluctuation) gradually. In the quantum annealing, in contrast, we decrease quantum field (quantum fluctuation) gradually and obtain a solution. In this paper we review how to implement quantum annealing and show some quantum fluctuation effects in frustrated Ising spin systems.

Keywords: Quantum annealing; Quantum fluctuation; Optimization problem; Transverse Ising model; Frustration.

1. Introduction

In information science, it has been an important issue to propose a novel method or improve known methods for obtaining the best solution of optimization problems. Optimization problem is to find the state in which the real-valued cost/gain function has the minimum/maximum value. In general, since there are huge number of elements in optimization problems, we cannot obtain the best solution by a naive method such as full search in limited time and present resources. This is because the number of states increases exponentially with the number of elements. Optimization problems are often represented by binary language of 0s and 1s. Here the minimum unit of information is called "bit". This situation is the same as the Ising model.[1] In the Ising model, the minimum unit corresponds to "spin" which can have the value $+1$ or -1. Then, in many cases, optimization problems can be represented by the Ising model with random interaction J_{ij}. The Hamiltonian is represented as

$$\mathcal{H}_{\text{Ising}} = -\sum_{i,j} J_{ij}\sigma_i^z \sigma_j^z, \quad \sigma_i^z = \pm 1. \tag{1}$$

Here σ_i^z is an assigned integer ± 1 (not operator[a]) at the i-th site. If an optimization problem is mapped onto the Ising Hamiltonian given by Eq. (1), the cost function of the optimization problem is expressed as the internal energy of the mapped Ising spin system. Thus, to find the best solution of optimization problem corresponds to obtain the ground state in the mapped Ising spin system.

Ising models with random interactions often have a difficulty of slow relaxation at low temperature. In general, energy landscape of complicated Ising model prevents the system from relaxation.[2–4] In order to avoid such a difficulty, many methods have been proposed in statistical physics. The most famous one is simulated annealing which was proposed by Kirkpatrick et al.[5,6] In the simulated annealing, we decrease temperature (thermal fluctuation) gradually and obtain a solution of optimization problem. In order to find the best solution, we should sample states according to transition probability defined in some way. Transition probability from a state Σ to the other state Σ' is often defined so as to be proportional to $e^{\beta(E(\Sigma)-E(\Sigma'))}$, where $E(\Sigma)$ is the eigenenergy of the eigenstate Σ and β is an inverse temperature $\beta = T^{-1}$. Here we set the Boltzmann constant k_B to be unity. At low temperature, because of large β, a transition between states does not occur with frequency. On the other hand, at high temperature, the probability distribution is almost flat. Then a transition from a certain state to the other state often occurs. By using the effect of thermal fluctuation, we are easy to obtain the best solution by decreasing temperature gradually. Actually, Kirkpatrick et al. demonstrated finding a stable state of the random Ising spin systems. After that, the simulated annealing is widely adopted for many kinds of optimization problems in physics, chemistry, and informa-

---

[a]Here we do not consider quantum system. In quantum systems $\hat{\sigma}_i^z$ represents the z-component of the Pauli matrix of the i-th site. Even if we consider the quantum version of the Hamiltonian given by Eq. (1), there is no off-diagonal elements. In this paper we discriminate c-number and matrix (operator) without or with the hat ˆ.

tion science since the simulated annealing is easy to implement. Moreover it has been proved that the simulated annealing can find the best solution definitely when we decrease temperature slow enough.[7] Then the simulated annealing has been regarded as guaranteed general method for optimization problems.

Since there are many types of optimization problems, there have been proposed corresponding methods in order to obtain the best solution of individual optimization problem. This is a style of information engineering. However it is also important to construct a general method for optimization problems such as the simulated annealing. Based on this point of view, the quantum annealing has been also developed.[8-30] In the quantum annealing, instead of decreasing temperature (thermal fluctuation), we decrease quantum field (quantum fluctuation) gradually and obtain a solution of optimization problem. The quantum annealing method has been expected as a powerful tool to obtain the best solution of optimization problem. However there are few examples that the efficiency of quantum annealing is worse than that of the simulated annealing. In order to know when to use the quantum annealing, it is a significant issue to investigate microscopic nature of quantum field response in simple models.

The organization of the rest of the paper is as follows. In Section 2, we review implementation methods of quantum annealing. In Section 3, we consider quantum field response of frustrated Ising spin systems. In Section 4, we conclude this paper briefly.

2. Implementation Methods of Quantum Annealing

In this section, we review concept of quantum annealing and implementation methods. The quantum annealing is expected to be an alternative method of simulated annealing and a general framework for optimization problems as mentioned above. Here we assume that our target optimization problem can be expressed by the Ising model given as

$$\mathcal{H}_c = -\sum_{i,j} J_{ij} \hat{\sigma}_i^z \hat{\sigma}_j^z, \qquad (2)$$

where $\hat{\sigma}_i^\alpha$ denotes the α-component of the Pauli matrix defined by

$$\hat{\sigma}_i^x = \begin{pmatrix} 0 & 1 \\ 1 & 0 \end{pmatrix}, \quad \hat{\sigma}_i^y = \begin{pmatrix} 0 & -i \\ i & 0 \end{pmatrix}, \quad \hat{\sigma}_i^z = \begin{pmatrix} 1 & 0 \\ 0 & -1 \end{pmatrix}. \qquad (3)$$

Equation (2) is a quantum version of the Hamiltonian given by Eq. (1) and a diagonal matrix. Our purpose is to find the ground state of the Hamiltonian given by Eq. (2). In order to perform quantum annealing, we introduce an additional time-dependent Hamiltonian $\mathcal{H}_q(t)$ which represents quantum fluctuation. Then the total Hamiltonian is expressed as

$$\mathcal{H}(t) = \mathcal{H}_c + \mathcal{H}_q(t). \tag{4}$$

Hereafter we refer to $\mathcal{H}_c$ and $\mathcal{H}_q(t)$ as classical Hamiltonian and quantum Hamiltonian, respectively. The most typical form of the quantum Hamiltonian for the Ising spin system is transverse field given as

$$\mathcal{H}_q(t) = -\Gamma(t) \sum_i \hat{\sigma}_i^x. \tag{5}$$

Although there is an ambiguity of $\mathcal{H}_q(t)$, we concentrate on the case that $\mathcal{H}_q(t)$ is the form given by Eq. (5) in this paper.

In the quantum annealing, we gradually decrease quantum field Γ and obtain the final state at $\Gamma = 0$, whereas we gradually decrease temperature T and obtain the final state at $T = 0$ in the simulated annealing (Fig. 1). Recently hybrid-type quantum simulated annealing in which we decrease temperature and quantum field simultaneously has developed.[19,20] Quantum annealing can be categorized by three types summarized in Table 1. There are some implementation methods of quantum annealing theoretically and experimentally. We can divide into two types of implementation theoretical methods according to how to treat time-development. The first one is a stochastic method which is realized by the Monte Carlo method.

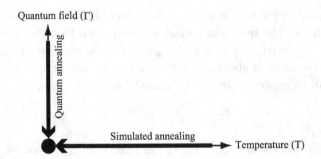

Fig. 1. Schematic diagram of simulated annealing and quantum annealing. The aim of annealing methods is to obtain the best solution at the circle ($T = \Gamma = 0$). In the simulated annealing, we decrease temperature T gradually from high temperature to (nearly) zero temperature. In the quantum annealing, we decrease quantum field Γ gradually from large value to (nearly) zero.

Table 1. Quantum annealing is categorized by three types depending on how to treat time-development.

Theoretical methods		Experiments
Stochastic methods	Deterministic methods	
Quantum Monte Carlo	Real-time dynamics (Exact diagonalization, time-dependent DMRG) Mean-field type method	Artificial lattices (Optical lattice)

The Monte Carlo methods have been widely adopted for analysis of equilibrium state of many-body strongly correlated systems in statistical physics. Since in some cases we face on the difficulty of obtaining the equilibrium state by naive Monte Carlo methods, many efficient algorithms have been developed.[21,31–37] Then if we adopt these engineered methods depending on the situation, we can use the Monte Carlo method as a powerful tool for quantum annealing.

The other type of theoretical methods is a deterministic methods. We can calculate real-time dynamics by solving the time-dependent Schrödinger equation directly[9–12,17,23] or performing the time-dependent density matrix renormalization group.[38] There is another type of deterministic method, which is mean-field type calculation.[39] Moreover, variational Bayes inference which is a useful method in information science can be regarded as a method based on mean-field calculation. This method can also treat large-scale problems as well as the Monte Carlo method. Originally, the variational Bayes inference has been adopted for many optimization problems. Recently a quantum annealing version of variational Bayes inference was developed.[20]

Quantum annealing can realize not only theoretically in classical computer but also experimentally. Artificial lattices such as optical lattice have been made in order to simulate strongly correlated fermionic systems and spin systems.[40–44] Since in artificial lattice systems, we can control parameters in the Hamiltonian, it is expected that these systems become a "quantum computer" which can find the ground state of complicated systems.

In this paper we focus on two theoretical implementation methods. In Sec. 2.1, we review the quantum annealing method for the transverse Ising model by using the quantum Monte Carlo method. In Sec. 2.2, we explain the real-time dynamics using the Schrödinger equation for the transverse Ising model.

2.1. Quantum Monte Carlo method

In this subsection, we explain how to implement quantum annealing by the quantum Monte Carlo method. Here we take the transverse Ising model as an example. The quantum Monte Carlo method is a useful tool to obtain thermodynamic properties of strongly correlated many-body systems such as quantum spin systems, bosonic systems, and fermionic systems. Before we review the quantum Monte Carlo method for the transverse Ising model, we show implementation method of classical Monte Carlo method for Ising spin systems without transverse field given by Eq. (1). The main procedure of the classical Monte Carlo method is as follows.

Step 1 We prepare an initial state.[b]
Step 2 We select a spin randomly.
Step 3 We flip the selected spin with probability defined by some way.
Step 4 We repeat Step 2 and Step 3 until physical quantities converge.

Performance of Monte Carlo method depends on how to choice the transition probability in Step 3. The simplest definition of the transition probability is the Metropolis method. Let Σ and Σ' be the state before and after spin flip, respectively. In the Metropolis method, we change the state from Σ to Σ' according to the following probability $P(\Sigma'|\Sigma)$:

$$P(\Sigma'|\Sigma) = \begin{cases} 1 & (E(\Sigma') \leq E(\Sigma)) \\ e^{\beta(E(\Sigma)-E(\Sigma'))} & (E(\Sigma') > E(\Sigma)) \end{cases}, \quad (6)$$

where $E(\Sigma)$ is the eigenenergy of the eigenstate Σ. In this rule, when the eigenenergy of the candidate state Σ' is lower than that of the present state, the state changes from $|\Sigma\rangle$ to $|\Sigma'\rangle$ definitely. Here $|\Sigma\rangle$ represents the direct product of N-spin states as follows:

$$|\Sigma\rangle = |\sigma_1\rangle \otimes |\sigma_2\rangle \otimes \cdots \otimes |\sigma_N\rangle, \quad (7)$$

where $|\sigma_i\rangle$ denotes the state at the i-th site. It should be noted that we can easily calculate $E(\Sigma)$ since now we consider the Hamiltonian in which there is no off-diagonal element.

Next we review the quantum Monte Carlo method for the transverse Ising model given by

$$\mathcal{H} = -\sum_{i,j} J_{ij} \hat{\sigma}_i^z \hat{\sigma}_j^z - \Gamma \sum_i \hat{\sigma}_i^x. \quad (8)$$

---

[b]There is an ambiguity how to prepare the initial state. For example, we can use both random spin configuration and completely polarized configuration.

In this model for large number of spins, it is difficult to obtain all eigenenergies and eigenstates in practice because of off-diagonal elements in the Hamiltonian. Then we cannot use the Monte Carlo method directly since we should know all eigenvalues and corresponding eigenstates. We have to consider an alternative representation which enables us to treat this model. In general, partition function of d-dimensional transverse Ising model (quantum system) is equivalent to that of $(d+1)$-dimensional Ising model without transverse field (classical system). This correspondence can be derived by the path-integral representation (in other words, the Trotter-Suzuki decomposition[45,46]).

Then we can use the Monte Carlo method for the transverse Ising model by considering path-integral representation. For simplicity, in order to demonstrate path-integral representation, we consider the ferromagnetic Ising chain with transverse field whose Hamiltonian is given by

$$\mathcal{H} = \mathcal{H}_c + \mathcal{H}_q, \quad \mathcal{H}_c = -J\sum_{i=1}^{N} \hat{\sigma}_i^z \hat{\sigma}_{i+1}^z, \quad \mathcal{H}_q = -\Gamma \sum_{i=1}^{N} \hat{\sigma}_i^x, \qquad (9)$$

where the periodic boundary is assumed: $\sigma_{N+1} = \sigma_1$. At the temperature $T(=\beta^{-1})$, the partition function of this system is expressed as

$$Z = \text{Tr}\, e^{-\beta \mathcal{H}} = \text{Tr}\, e^{-\beta(\mathcal{H}_c + \mathcal{H}_q)} = \sum_{\Sigma} \left\langle \Sigma \left| e^{-\beta(\mathcal{H}_c + \mathcal{H}_q)} \right| \Sigma \right\rangle. \qquad (10)$$

We cannot obtain the partition function by this direct expression since $\langle \Sigma | e^{-\beta(\mathcal{H}_c + \mathcal{H}_q)} | \Sigma \rangle$ is not tractable. Then by using integer m, we decompose the matrix exponential $e^{-\beta(\mathcal{H}_c + \mathcal{H}_q)}$ as follows.

$$\exp\left[-\beta(\mathcal{H}_c + \mathcal{H}_q)\right] = \exp\left[e^{-\frac{1}{m}\beta \mathcal{H}_c} e^{-\frac{1}{m}\beta \mathcal{H}_q}\right]^m + \mathcal{O}\left(\left(\frac{\beta}{m}\right)^2\right). \qquad (11)$$

Then we obtain the partition function by using m:

$$\begin{aligned} Z &= \sum_{\Sigma} \left\langle \Sigma \left| e^{-\beta(\mathcal{H}_c + \mathcal{H}_q)} \right| \Sigma \right\rangle \\ &= \sum_{\Sigma_1, \Sigma_1', \cdots, \Sigma_m, \Sigma_m'} \left\langle \Sigma_1 \left| e^{-\frac{\beta}{m}\mathcal{H}_c} \right| \Sigma_1' \right\rangle \left\langle \Sigma_1' \left| e^{-\frac{\beta}{m}\mathcal{H}_q} \right| \Sigma_2 \right\rangle \\ &\quad \times \left\langle \Sigma_2 \left| e^{-\frac{\beta}{m}\mathcal{H}_c} \right| \Sigma_2' \right\rangle \left\langle \Sigma_2' \left| e^{-\frac{\beta}{m}\mathcal{H}_q} \right| \Sigma_3 \right\rangle \\ &\quad \times \cdots \\ &\quad \times \left\langle \Sigma_m \left| e^{-\frac{\beta}{m}\mathcal{H}_c} \right| \Sigma_m' \right\rangle \left\langle \Sigma_m' \left| e^{-\frac{\beta}{m}\mathcal{H}_q} \right| \Sigma_1 \right\rangle, \end{aligned} \qquad (12)$$

where $|\Sigma_k\rangle$ represents the direct product of N spin states as well as $|\Sigma\rangle$:

$$|\Sigma_k\rangle = |\sigma_{k,1}\rangle \otimes |\sigma_{k,2}\rangle \otimes \cdots \otimes |\sigma_{k,N}\rangle. \qquad (13)$$

The index k represents the coordinate along the Trotter axis. Hereafter we refer to m as the Trotter number. Since the classical Hamiltonian $\mathcal{H}_c$ is a diagonal matrix, the following relation is satisfied:

$$\hat{\sigma}_j^z |\Sigma_k\rangle = \sigma_{k,j}^z |\Sigma_k\rangle. \quad (14)$$

From the above relation and simple calculation,

$$\left\langle \Sigma_k \left| e^{-\frac{\beta \mathcal{H}_c}{m}} \right| \Sigma_k' \right\rangle = \exp\left(\frac{\beta J}{m} \sum_{i=1}^{N} \sigma_{k,i}^z \sigma_{k,i+1}^z\right) \prod_{i=1}^{N} \delta(\sigma_{k,i}^z, \sigma_{k,i}^{z'}),$$

$$\left\langle \Sigma_k' \left| e^{-\frac{\beta \mathcal{H}_q}{m}} \right| \Sigma_{k+1} \right\rangle = \left[\frac{1}{2} \sinh\left(\frac{2\beta\Gamma}{m}\right)\right]^{\frac{N}{2}} \exp\left[\frac{1}{2} \log \coth\left(\frac{\beta\Gamma}{m}\right) \sum_{i=1}^{N} \sigma_{k,i}^{z'} \sigma_{k+1,i}^z\right],$$

are obtained. Then the partition function is expressed by the following way:

$$Z = \lim_{m \to \infty} \left[\frac{1}{2} \sinh\left(\frac{2\beta\Gamma}{m}\right)\right]^{\frac{N}{2}}$$

$$\times \sum_{\{\sigma_{k,i}=\pm 1\}} \exp\left[\sum_{i=1}^{N}\sum_{k=1}^{m} \left(\frac{\beta J}{m}\sigma_{k,i}^z \sigma_{k,i+1}^z + \frac{1}{2}\log\coth\left(\frac{\beta\Gamma}{m}\right)\sigma_{k,i}^z \sigma_{k+1,i}^z\right)\right]$$

$$= \lim_{m \to \infty} A \sum_{\{\sigma_{k,i}=\pm 1\}} \exp(-\beta\mathcal{H}_{\text{eff}}), \quad (15)$$

where A is just a coefficient which is irrelevant for thermodynamic properties and the effective Hamiltonian $\mathcal{H}_{\text{eff}}$ is defined as

$$\mathcal{H}_{\text{eff}} = \sum_{i=1}^{N}\sum_{k=1}^{m} \left[-\frac{J}{m}\sigma_{k,i}^z\sigma_{k,i+1}^z - \frac{1}{2\beta}\log\coth\left(\frac{\beta\Gamma}{m}\right)\sigma_{k,i}^z\sigma_{k+1,i}^z\right]. \quad (16)$$

This relation means the partition function of one-dimensional transverse Ising model is equivalent to that of two-dimensional Ising model without transverse field. Since physical quantities can be calculated by the partition function in general, we can obtain thermodynamic properties of transverse Ising chain by considering that of two-dimensional classical Ising model. It should be noted that the abovementioned derivation does not depend on spatial dimension. Then the method can be adopted for general Ising model with transverse field. Figure 2 displays schematic spin states of the transverse Ising chain for small Γ and large Γ at low temperature. When the transverse field Γ is small, spins are almost aligned along both real space and the Trotter axis. On the other hand, spins are aligned along only real space when the transverse field Γ is large. This is a nature which comes from quantum fluctuation effect.

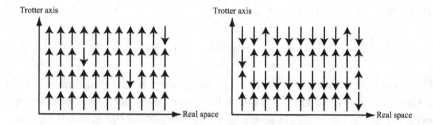

Fig. 2. Schematic spin configurations of the transverse Ising chain given by Eq.(9) for small Γ (left panel) and large Γ (right panel).

So far, we review quantum Monte Carlo method as a tool to obtain the thermodynamic equilibrium properties. This method can be also regarded as a realization method of stochastic dynamics. Then we can use quantum Monte Carlo method as a method to perform quantum annealing. In the quantum annealing, the transverse field decreases against Monte Carlo step. The efficiency of quantum annealing by quantum Monte Carlo method has been studied, and the quantum annealing method has been succeeded to obtain not so bad solution of some optimization problems.

2.2. Real-time dynamics

In the previous subsection, we explained implementation method of quantum annealing by quantum Monte Carlo method. In the method, time-evolution is treated as a stochastic process. In this subsection, we review a method in which time-evolution is considered as a deterministic process. There are a couple of methods which realize deterministic time-evolution. In this paper, we focus on real-time dynamics which is derived from the Schrödinger equation:

$$i\hbar \frac{\partial}{\partial t} |\psi(t)\rangle = \mathcal{H}(t) |\psi(t)\rangle, \qquad (17)$$

where $|\psi(t)\rangle$ and $\mathcal{H}(t)$ denote time-dependent state and Hamiltonian, respectively, at time t. In this method, once we prepare a state as the initial state, the time-development of the state is determined. Then, how to choice the initial state is important. In the sense of quantum annealing, it is a usual way to prepare the initial state which can be made easily. In the case of the transverse Ising model, fully polarized state is a trivial state at large transverse field. The state is expressed as

$$|\psi(0)\rangle = |\rightarrow \cdots \rightarrow\rangle, \qquad (18)$$

where the state $|\rightarrow\rangle$ is defined as

$$|\rightarrow\rangle = \frac{1}{\sqrt{2}}(|\uparrow\rangle + |\downarrow\rangle). \qquad (19)$$

The state $|\rightarrow\rangle$ is an eigenstate of the x-component of the Pauli matrix $\hat{\sigma}^x$. If the energy level of the prepared initial state does not cross with other levels for all transverse field strength Γ, we can obtain the ground state in the adiabatic limit. However, in practice, since we decrease the transverse field with finite speed, transition to excited levels is inevitable. In order to consider such a nonadiabatic transition, we study a single spin system with longitudinal and transverse fields. The Hamiltonian is given by

$$\mathcal{H}_{\text{single}} = -h\hat{\sigma}^z - \Gamma\hat{\sigma}^x = \begin{pmatrix} -h & -\Gamma \\ -\Gamma & h \end{pmatrix}. \qquad (20)$$

Eigenvalues of this Hamiltonian are

$$E_\pm = \pm\sqrt{h^2 + \Gamma^2}. \qquad (21)$$

Figure 3 denotes the eigenenergies as functions of longitudinal field h. Then the energy difference (energy gap) between two eigenstates is $2\sqrt{h^2 + \Gamma^2}$. The energy gap takes the minimum value 2Γ at $h = 0$ for Γ.

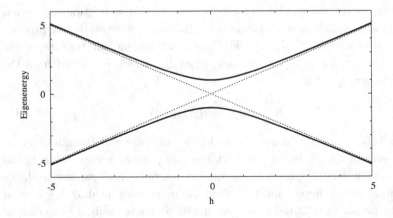

Fig. 3. Eigenenergies of the single spin with longitudinal and transverse magnetic fields. The dotted lines and the solid curves indicate the eigenenergies for $\Gamma = 0$ and $\Gamma = 1$, respectively.

Next we consider a single spin problem with time-dependent longitudinal field $h(t) = vt$ and the fixed transverse field Γ:

$$\mathcal{H}_{\text{single}}(t) = -vt\hat{\sigma}^z - \Gamma\hat{\sigma}^x = \begin{pmatrix} -vt & -\Gamma \\ -\Gamma & vt \end{pmatrix}. \tag{22}$$

Here we set the down state $|\downarrow\rangle$ as the initial state. This state is the ground state of the Hamiltonian in the limit of $t \to -\infty$. When the transverse field Γ is zero, the state $|\downarrow\rangle$ is the eigenstate of the Hamiltonian. From the viewpoint of quantum annealing, if the symmetry of the prepared initial state is different from that of the ground state at the final time, we cannot obtain the ground state at all. We next consider the case for finite transverse field Γ. In this case, the energy level of the ground state does not cross with that of the excited state as stated before. Then the state becomes $|\uparrow\rangle$ at $t \to \infty$ in the adiabatic limit ($v \to 0$). When we sweep longitudinal magnetic field with finite speed v, linear combination of eigenstates is obtained even at $t \to \infty$.

The left and right panels in Fig. 4 show time-development of magnetization along the z-axis for various sweeping speed with fixed transverse field $\Gamma = 0.2$ and for various Γ with fixed sweeping speed $v = 0.1$, respectively. Here the magnetization along the z-axis is calculated by

$$m^z(t) = \langle \psi(t)|\hat{\sigma}^z|\psi(t)\rangle. \tag{23}$$

As the sweeping speed becomes slow and/or the energy gap becomes large, the state approaches the adiabatic limit of the ground state ($m^z(t = +\infty) = +1$). This is called the Landau-Zener-Stückelberg transition.[47–49] The asymptotic behavior of the transition probability at $t \to +\infty$ is given

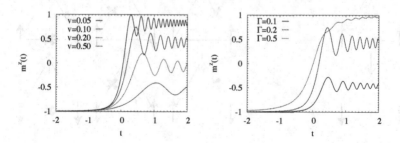

Fig. 4. Time-development of the magnetization along the z-axis for various sweeping speed under fixed $\Gamma = 0.2$ (left panel) and for various Γ under fixed sweeping speed $v = 0.1$ (right panel).

by

$$P_{\text{LZS}} = \exp\left(-\frac{(\Delta E)^2}{4v}\right), \qquad (24)$$

where ΔE is the energy gap at $h = 0$; in this case $\Delta E = 2\Gamma$. The Landau-Zener-Stückelberg transition is adopted not only for single spin problem but also many-body quantum systems. Actually, a couple of quantum dynamical behaviors can be analyzed by the Landau-Zener-Stückelberg transition since in many cases, the energy level structure can be approximated by that of single spin system. Then, in order to consider nonadiabatic transition in the quantum annealing, the knowledge of the Landau-Zener-Stückelberg transition is useful.

3. Quantum Field Response of Frustrated Ising Systems

In this section we consider transverse field response of frustrated Ising spin systems. In order to explain the concept of frustration, we first consider the Ising model on a triangle cluster. The Hamiltonian of this model is given by

$$\mathcal{H}_\triangle = -J\left(\sigma_1^z \sigma_2^z + \sigma_2^z \sigma_3^z + \sigma_3^z \sigma_1^z\right), \qquad \sigma_i^z = \pm 1. \qquad (25)$$

When the coupling constant J is positive/negative, this is called ferromagnetic/antiferromagnetic interaction. Figure 5 (a) shows the ground states in the case of ferromagnetic interaction. In the ground states, all interactions are energetically favorable. On the other hand, when the interactions are antiferromagnetic ($J < 0$), in the ground state, there is a bond which

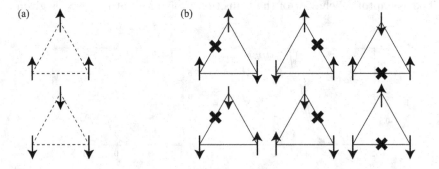

Fig. 5. Ground state spin configuration of the Ising model on a triangle cluster. The dotted and solid lines represent ferromagnetic and antiferromagnetic interactions, respectively. (a) Ferromagnetic case. All interactions are energetically favorable states. (b) Antiferromagnetic case. The crosses denote energetically unfavorable states.

is energetically unfavorable shown in Fig. 5(b). Such a nature is called frustration. Since frustration prevents the system from ordering and makes peculiar density of states, unconventional order and characteristic dynamic behaviors appear.[24,50-60]

Frustration appears in antiferromagnet on triangle-based lattices such as triangular lattice, kagomé lattice, and pyrochlore lattice. In random Ising spin systems, frustration also appears randomly on the lattice. Since many optimization problems can be mapped onto random Ising models as mentioned above, it is an important issue to investigate quantum field response of frustrated systems comparing with thermal fluctuation effect.

In this paper we focus on two examples of frustrated Ising spin systems. In Sec. 3.1, we review transverse field effect on fully frustrated systems. In Sec. 3.2, we study decorated bond systems in which correlation function behaves nonmonotonic against temperature and transverse magnetic field.

3.1. *Order by disorder effect in fully frustrated systems*

Here we consider fully frustrated systems where all plaquettes are frustrated. There are macroscopically degenerated ground states in the fully frustrated systems.[62-67] Typical configurations of ground state of the antiferromagnetic system on the triangular lattice are shown in Fig. 6. In Fig. 6, we adopt the periodic boundary condition. The dotted boxes in these figures indicate "free spin" where the internal field from the nearest neighbor sites is zero. The number of free spins is a useful way to represent a character of each configuration. From left to right in Fig. 6, the number of free spins decreases. Here we first consider simulated annealing for such fully frustrated systems. From the principle of equal weight, all distinct degenerated ground states can be obtained with the same probability at $T = 0+$ in the simulated annealing.

By the way when we adopt quantum annealing for fully frustrated systems, whether does the probability distribution of the obtained ground states become flat or not? If the probability distribution becomes biased distribution, which are states selected? From the preceding studies, it is well-known that the states which have many free spins tend to appear in the quantum annealing.[25,28,30,61] The left panel in Fig. 6 represents the maximum free spin state. Actually the probability of this state is the maximum in the adiabatic limit. This is because the transverse field is represented as

$$-\Gamma \sum_i \hat{\sigma}_i^x = -\Gamma \sum_i \left(\hat{\sigma}_i^+ + \hat{\sigma}_i^- \right), \qquad (26)$$

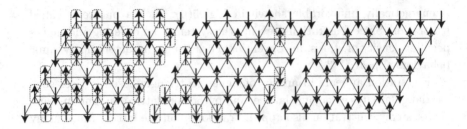

Fig. 6. Typical ground states of antiferromagnetic Ising spin system on triangular lattice. The dotted boxes represent free spin where the internal field from the nearest neighbor sites are zero.

and corresponds to sum of the spin-flip operator. Then the states which have large number of free spins appear with high probability. Schematic picture of the probability distributions of the ground state obtained by the simulated annealing and the quantum annealing is shown in Fig. 7. The maximum free spin states are "ordered states" which are mediated by quantum fluctuation effect. Then, the nature is called "order by disorder" which is a famous feature in fully frustrated systems.[51-53] The energy of the states obtained by simulated annealing with slow schedule is the same as that of the states obtained by quantum annealing. Although the maximum free spin states are selected in this case, the other type of ground states can be selected when we adopt the other type of quantum fluctuation. It is an interesting problem to investigate such probability distribution when we decrease temperature and transverse field with finite speed.[68]

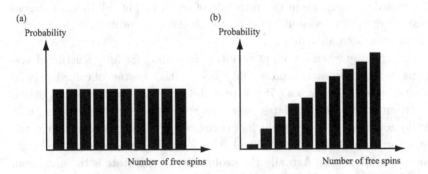

Fig. 7. Schematic picture of probability distributions of the ground state for (a) simulated annealing and (b) quantum annealing. The number of free spins represents the configuration of ground states.

3.2. Decorated bond systems

Next we consider quantum field response of decorated bond systems. Figure 8 shows a structure of decorated bond systems. The circles and triangles in Fig. 8 denote system spins and decorated spins, respectively. Before we consider thermodynamic properties of lattice system with decorated bond shown in Fig. 8 (b), we first consider temperature dependent correlation function between system spins of the decorated bond unit shown in Fig. 8 (a). The Hamiltonian of the unit is given as

$$\mathcal{H}_{\text{unit}} = -J_{\text{dir}}\sigma_1^z \sigma_2^z - J \sum_{i=1}^{N_d} (\sigma_1^z + \sigma_2^z) s_i^z, \qquad \sigma_i^z = \pm 1, s_i^z = \pm 1, \quad (27)$$

where J_{dir} and J represent direct coupling between system spins and decorated bond between a system spin and a decorated spin, respectively.

The correlation function between system spins can be calculated exactly as

$$\langle \sigma_1^z \sigma_2^z \rangle = \tanh K_{\text{eff}}, \qquad (28)$$

$$\text{Tr}_{\{s_i\}} e^{-\beta \mathcal{H}} = A e^{K_{\text{eff}} \sigma_1^z \sigma_2^z}, \qquad (29)$$

$$K_{\text{eff}} = \frac{N_d}{2} \log \cosh(2\beta J) + \beta J_{\text{dir}}, \qquad (30)$$

where A is an irrelevant factor. Here if $J_{\text{dir}} = 0$ and $J > 0$, the correlation function is always positive and monotonic decreasing function against temperature. On the other hand, when $J = 0$ and $J_{\text{dir}} < 0$, the correlation function is always negative and monotonic increasing function against temperature. Then, by tuning the ratio J_{dir} and J with keeping $J_{\text{dir}} < 0$ and

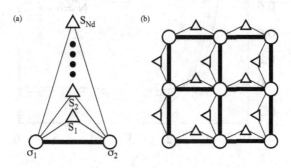

Fig. 8. (a) Unit of the decorated bond system. (b) Square lattice with decorated bond for $N_d = 1$.

$J > 0$, the correlation function $\langle \sigma_1^z \sigma_2^z \rangle$ behaves non-monotonic as a function of temperature.[24,54,69–73] In this paper we adopt $J_{\text{dir}} = -\frac{N_d}{2} J$ and J as an energy unit. The temperature dependency of correlation function in this case is shown in Fig. 9. It is noted that the signs of correlation function and the effective coupling K_{eff} have both plus and minus values depending on temperature.

In order to investigate the reason for such a non-monotonic behavior, we consider probability distributions of ferromagnetically correlated states ($\sigma_1^z \sigma_2^z = +1$) and antiferromagnetically correlated states ($\sigma_1^z \sigma_2^z = -1$). The number of ferromagnetically correlated states and that of antiferromagnetically correlated states are 2^{N_d+1}. Let N_s be the number of decorated spins in which directions of these spins are the same as σ_1^z. The probability distributions of ferromagnetically correlated states $P_F(N_s)$ and antiferromagnetically correlated states $P_{AF}(N_s)$ are given by

$$P_F(N_s) = \frac{e^{-\beta \frac{N_d}{2} J}}{2 \cosh\left(\beta \frac{N_d}{2} J\right)} \binom{N_d}{N_s} \left(\frac{1}{2}\right)^{N_d} \frac{e^{-2\beta(N_d - 2N_s)J}}{\cosh^{N_d}(2\beta J)}, \quad (31)$$

$$P_{AF}(N_s) = \frac{e^{\beta \frac{N_d}{2} J}}{2 \cosh\left(\beta \frac{N_d}{2} J\right)} \binom{N_d}{N_s} \left(\frac{1}{2}\right)^{N_d}. \quad (32)$$

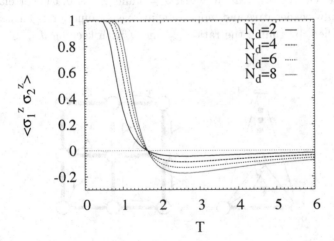

Fig. 9. Correlation function between system spins as a function of temperature.

It should be noted that the following relation is obviously satisfied:

$$\sum_{N_s=0}^{N_d} [P_F(N_s) + P_{AF}(N_s)] = 1. \qquad (33)$$

At zero temperature all system spins and decorated spins are the same value. Then the probability distribution is as follows: $P_F(N_d) = 1$ and otherwise are zero. At the temperature where $\sum_{N_s} P_F(N_s) = \sum_{N_s} P_{AF}(N_s)$ is satisfied, the correlation function $\langle \sigma_1^z \sigma_2^z \rangle$ becomes zero. This is the reason why the correlation function and effective coupling behave non-monotonic.

Suppose we consider square lattice system with decorated bond for large enough N_d. In this case there is temperature region where the absolute value of effective coupling exceeds the critical value of the Ising spin system on square lattice $K_c = \frac{1}{2}\log(1+\sqrt{2})$.[74] Then, paramagnetic phase → antiferromagnetic phase → paramagnetic phase → ferromagnetic phase appear as temperature decreases and at each phase boundary, phase transition takes place. Such successive phase transitions are called reentrant phase transition. Reentrant phase transitions and non-monotonic behavior of correlation function are typical nature of frustrated systems.[75]

So far we showed thermal fluctuation response of decorated Ising spin systems. Next we consider transverse field response of this system at zero temperature. Then the Hamiltonian is given by

$$\mathcal{H} = -J_{\text{dir}} \hat{\sigma}_1^z \hat{\sigma}_2^z - J \sum_{i=1}^{N_d} (\hat{\sigma}_1^z + \hat{\sigma}_2^z) \hat{s}_i^z - \Gamma \left(\hat{s}_1^x + \hat{s}_2^x + \sum_{i=1}^{N_d} \hat{\sigma}_i^x \right), \qquad (34)$$

where $\hat{s}_i^\alpha$ denotes the α-element of the Pauli matrix of the i-th decorated spin. As the previous example, we adopt $J_{\text{dir}} = -\frac{N_d}{2}J$. We calculate the correlation function between system spins along the z-axis at zero temperature as a function of transverse field as shown in Fig. 10. As well as the thermal fluctuation, the correlation function behaves non-monotonic as a function of transverse field. This result indicates there is a similar point between thermal fluctuation and quantum fluctuation originated by transverse field. Suppose we consider the system where reentrant phase transition occurs when we decrease temperature. If we adopt the simulated annealing method, we face on the difficulty which comes from the critical slowing down and the simulated annealing is not useful. Whenever we adopt the quantum annealing using time-dependent transverse field, the system also exhibits reentrant phase transition. Then we should consider other type of quantum fluctuation effect in order to avoid the occurrence of the phase transition.

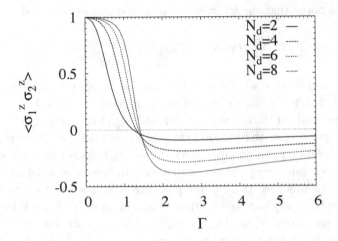

Fig. 10. Correlation function between the system spins as a function of transverse field.

4. Conclusion and Future Perspective

In this paper, we reviewed how to implement quantum annealing focusing on the methods based on quantum Monte Carlo method and time-development Schrödinger equation. The quantum annealing method is expected to be a powerful tool to obtain the best solution of optimization problems. This method is a general method and can be implemented easily. In this method we use quantum fluctuation to find the stable state. In order to put into practical use, it is important to investigate quantum fluctuation effects for optimization problems. Then we considered quantum fluctuation effect for frustrated systems by taking the transverse Ising model as an example, since the Ising model which represents optimization problem is a frustrated system in general. We explained differences and similarities between thermal fluctuation and quantum fluctuation. By building up the organized knowledge of the thermal fluctuation and the quantum fluctuation, the quantum annealing method develops into a truly useful algorithm.

Acknowledgement

The authors are grateful to Bernard Barbara, Masaki Hirano, Yoshiki Matsuda, Seiji Miyashita, Hans de Raedt, Per Arne Rikvold, and Eric Vincent for their valuable comments. R.T. is partly supported by Global COE Program "the Physical Sciences Frontier", MEXT, Japan. S.T. is partly supported by Grand-in-Aid for JSPS Fellows (23-7601). The computation in the present work was performed on computers at the Supercomputer Center, Institute for Solid State Physics, University of Tokyo.

References

1. E. Ising, *Z. Physik* **31**, 253 (1925).
2. M. Mézard, G. Parisi, and M. A. Virasoro, *Spin Glass Theory and Beyond* (World Scientific, 1987).
3. K. H. Fischer and J. A. Hertz, *Spin Glasses* (Cambridge University Press, 1993).
4. A. P. Young, *Spin Glasses and Random Fields* (World Scientific, 1998).
5. S. Kirkpatrick, C. D. Gelatt Jr., and M. P. Vecchi, *Science* **220**, 671 (1983).
6. S. Kirkpatrick, *J. Stat. Phys.* **34**, 975 (1984).
7. S. Geman and D. Geman, *IEEE Transactions on Pattern Analysis and Machine Intelligence* **6**, 721 (1984).
8. A. B. Finnila, M. A. Gomez, C. Sebenik, C. Stenson, and J. D. Doll, *Chem. Phys. Lett.* **219** 343 (1994).
9. T. Kadowaki and H. Nishimori, *Phys. Rev. E* **58**, 5355 (1998).
10. J. Brooke, D. Bitko, T. F. Rosenbaum, and G. Aeppli, *Science* **284**, 779 (1999).
11. E. Farhi, J. Goldstone, S. Gutmann, J. Lapan, A. Lundgren, and D. Preda, *Science* **292**, 472 (2001).
12. G. E. Santoro, R. Martoňák, E. Tosatti, and R. Car, *Science* **295**, 2427 (2002).
13. R. Martoňák, G. E. Santoro, and E. Tosatti, *Phys. Rev. E* **70**, 057701 (2004).
14. D. A. Battaglia, G. E. Santoro, and E. Tosatti, *Phys. Rev. E* **71**, 066707 (2005).
15. S. Suzuki and M. Okada, *J. Phys. Soc. Jpn.* **74**, 1649 (2005).
16. A. Das and B. K. Chakrabarti, *Quantum Annealing and Related Optimization Methods* (Springer, 2005).
17. S. Tanaka and S. Miyashita, *J. Magn. Magn. Mater.* **310**, e468 (2007).
18. A. Das and B. K. Chakrabarti, *Rev. Mod. Phys.* **80**, 1061 (2008).
19. K. Kurihara, S. Tanaka, and S. Miyashita, *Proceedings of the 25th Conference on Uncertainty in Artificial Intelligence* (2009).
20. I. Sato, K. Kurihara, S. Tanaka, H. Nakagawa, and S. Miyashita, *Proceedings of the 25th Conference on Uncertainty in Artificial Intelligence* (2009).
21. S. Morita, S. Suzuki, and T. Nakamura, *Phys. Rev. E* **79**, 065701(R) (2009).
22. J. Inoue, Y. Saika, and M. Okada, *Lecture Note in Physics "Quantum Quenching, Annealing, and Computation"* (Springer) **802**, 283 (2010).
23. S. Miyashita, S. Tanaka, H. de Raedt, and B. Barbara, *J. Phys.: Conf. Ser.* **143**, 012005 (2009).
24. S. Tanaka and S. Miyashita, *Phys. Rev. E* **81**, 051138 (2010).
25. S. Tanaka, M. Hirano, and S. Miyashita, *Lecture Note in Physics "Quantum Quenching, Annealing, and Computation"* (Springer) **802**, 215 (2010).
26. A. K. Chandra, A. Das, J. Inoue, and B. K. Chakrabarti, *Lecture Note in Physics "Quantum Quenching, Annealing, and Computation"* (Springer) **802**, 235 (2010).

27. M. Ohzeki and H. Nishimori, *J. Comp. and Theor. Nanoscience* **8**, 963 (2011).
28. S. Tanaka, M. Hirano, and S. Miyashita, *Physica E* **43**, 766 (2010).
29. S. Tanaka, R. Tamura, I. Sato, and K. Kurihara, to appear in *Kinki University Quantum Computing Series: "Summer School on Diversities in Quantum Computation/Information"*.
30. S. Tanaka, to appear in *proceedings of Kinki University Quantum Computing Series: "Symposium on Quantum Information and Quantum Computing"* (2011).
31. R. H. Swendsen and J. S. Wang, *Phys. Rev. Lett.* **58**, 86 (1987).
32. U. Wolff, *Phys. Rev. Lett.* **62**, 361 (1989).
33. K. Hukushima and K. Nemoto, *J. Phys. Soc. Jpn.* **65**, 1604 (1996).
34. N. Kawashima and K. Harada, *J. Phys. Soc. Jpn.* **73**, 1379 (2004).
35. T. Nakamura, *Phys. Rev. Lett.* **101**, 210602 (2008).
36. H. Suwa and S. Todo, *Phys. Rev. Lett.* **105**, 120603 (2010).
37. H. Suwa and S. Todo, *arXiv*:1106.3562.
38. S. Suzuki and M. Okada, *Interdisciplinary Information Sciences* **13**, 49 (2007).
39. K. Tanaka and T. Horiguchi, *Electronics and Communications in Japan* **83**, 84 (2000).
40. I. Bloch, J. Dalibard, W. Zwerger, *Rev. Mod. Phys.* **80**, 885 (2008).
41. K. Kim, M. -S. Chang, S. Korenblit, R. Islam, E. E. Edwards, J. K. Freericks, G. -D. Lin, L. -M. Duan, and C. Monroe, *Nature* **465**, 590 (2010).
42. X. -S. Ma, B. Dakic, W. Naylor, A. Zeilinger, and P. Walther, *Nat. Phys.* **7**, 399 (2011).
43. J. Simon, W. S. Bakr, R. Ma, M. E. Tai, P. M. Preiss, and M. Greiner, *Nature* **472**, 307 (2011).
44. J. Struck, C. Ölschläger, R. L. Targat, P. S. Panahi, A. Eckardt, M. Lewenstein, P. Windpassinger, and K. Sengstock, *Science* **333**, 996 (2011).
45. H. F. Trotter, *Proceedings of the American Mathematical Society* **10**, 545 (1959).
46. M. Suzuki, *Prog. Theor. Phys.* **56**, 1454 (1976).
47. L. Landau, *Phys. Z. Souwjetunion* **2**, 46 (1932).
48. C. Zener, *Proc. R. Soc. London Ser. A* **137**, 696 (1932).
49. E. C. G. Stückelberg, *Helv. Phys. Acta* **5**, 369 (1932).
50. G. Toulouse, *Commun. Phys.* (London) **2**, 115 (1977).
51. R. Liebmann, *Statistical Mechanics of Periodic Frustrated Ising Systems* (Springer-Verlag, Berlin/Heidelberg, GmbH, Heidelberg, 1986).
52. H. Kawamura, *J. Phys.: Condens. Matter* **10**, 4707 (1998).
53. H. T. Diep (ed.), *Frustrated Spin Systems* (World Scientific, Singapore, 2005).
54. S. Tanaka and S. Miyashita, *Prog. Theor. Phys. Suppl.* **157**, 34 (2005).
55. S. Miyashita, S. Tanaka, and M. Hirano, *J. Phys. Soc. Jpn.* **76**, 083001 (2007).
56. S. Tanaka and S. Miyashita, *J. Phys. Soc. Jpn.* **76**, 103001 (2007).
57. R. Tamura and N. Kawashima, *J. Phys. Soc. Jpn.* **77**, 103002 (2008).
58. S. Tanaka and S. Miyashita, *J. Phys. Soc. Jpn.* **78**, 084002 (2009).
59. R. Tamura and N. Kawashima, *J. Phys. Soc. Jpn.* **80**, 074008 (2011).
60. R. Tamura, N. Kawashima, T. Yamamoto, C. Tassel, and H. Kageyama, *Phys. Rev. B* **84**, 214408 (2011).

61. Y. Matsuda, H. Nishimori, and H. G. Katzgraber, *New J. Phys.* **11**, 073021 (2009).
62. K. Husimi and I. Syozi, *Prog. Theor. Phys.* **5**, 177 (1949).
63. I. Syozi, *Prog. Theor. Phys.* **5**, 341 (1949).
64. R. M. F. Houtappel, *Physica* **16**, 425 (1950).
65. G. H. Wannier, *Phys. Rev.* **79**, 357 (1950).
66. K. Kano and S. Naya, *Prog. Theor. Phys.* **10**, 158 (1953).
67. G. H. Wannier, *Phys. Rev. B* **7**, 5017 (1973).
68. S. Tanaka and R. Tamura, *in preparation*.
69. E. H. Fradkin and T. P. Eggarter, *Phys. Rev. A* **14**, 495 (1976).
70. S. Miyashita, *Prog. Theor. Phys.* **69**, 714 (1983).
71. H. Kitatani, S. Miyashita, and M. Suzuki, *Phys. Lett.* **108A**, 45 (1985).
72. H. Kitatani, S. Miyashita, and M. Suzuki, *J. Phys. Soc. Jpn.* **55**, 865 (1986).
73. S. Miyashita and E. Vincent, *Eur. Phys. J. B* **22**, 203 (2001).
74. L. Onsager, *Phys. Rev.* **65**, 117 (1944).
75. P. Azaria, H. T. Diep, and H. Giacomini, *Phys. Rev. Lett.* **59**, 1629 (1987).